An Updated Look at Military and Civilian Pay Levels and Recruit Quality

TROY D. SMITH, BETH J. ASCH, MICHAEL G. MATTOCK

Prepared for the Office of the Secretary of Defense
Approved for public release; distribution unlimited

NATIONAL DEFENSE RESEARCH INSTITUTE

For more information on this publication, visit www.rand.org/t/RR3254

Library of Congress Cataloging-in-Publication Data is available for this publication.
ISBN: 978-1-9774-0393-3

Published by the RAND Corporation, Santa Monica, Calif.
© Copyright 2020 RAND Corporation
RAND® is a registered trademark.

Cover: U.S. Marine Corps photo by Lance Cpl. Phuchung Nguyen.

Support RAND
Make a tax-deductible charitable contribution at
www.rand.org/giving/contribute

www.rand.org

Preface

Quadrennial reviews of military compensation seek to ensure that pay and benefit levels for those serving in the military are adequate and able to attract the quality and quantity of recruits necessary to maintain readiness. This report, in support of the 13th Quadrennial Review of Military Compensation, builds on earlier RAND work (Hosek et al., 2018) by examining the current state of military compensation relative to civilian pay for workers of comparable ages, education levels, and labor-force participation. The Ninth Quadrennial Review of Military Compensation recommended that military pay for active-component enlisted personnel be at about the 70th percentile of civilian pay for full-time workers with some college and that military pay for active-component officers be at about the 70th percentile of civilian pay for full-time workers with four or more years of college. We compare relative pay for enlisted and officers in 2017 with their relative pay in 2009. We also examine how changes in military pay affect the quality of recruits across branches of the military, as well as how pay percentiles vary by geography.

The current research was sponsored by the 13th Quadrennial Review of Military Compensation and conducted within the Forces and Resources Policy Center of the RAND National Defense Research Institute, a federally funded research and development center sponsored by the Office of the Secretary of Defense, the Joint Staff, the Unified Combatant Commands, the Navy, the Marine Corps, the defense agencies, and the defense Intelligence Community.

For more information on the RAND Forces and Resources Policy Center, see www.rand.org/nsrd/ndri/centers/frp or contact the director (contact information is provided on the webpage).

Contents

Figures

Tables

Summary

Since the beginning of the all-volunteer force in the early 1970s, compensation and benefits have been among the most critical tools for attracting and retaining the quantity and quality of military personnel necessary for the United States to achieve its military goals. Military pay must be high enough to attract quality recruits away from other jobs that they could get given their education, skills, and ability while also being respectful of the public trust to appropriately manage public funds.[1] The Ninth Quadrennial Review of Military Compensation (QRMC) measured military pay via regular military compensation (RMC), which is the sum of basic pay, basic allowance for housing (BAH), basic allowance for subsistence (BAS), and the federal tax advantage resulting from allowances not being taxed. The report concluded, "Pay at around the 70th percentile of comparably educated civilians has been necessary to enable the military to recruit and retain the quantity and quality of personnel it requires" (Office of the Under Secretary of Defense for Personnel and Readiness, 2002, p. xxiii).[2]

---

[1] This report focuses on active-component (AC) compensation and does not directly consider reserve-component (RC) compensation, though elements of the RC system are tied to that of the AC system, such as the pay table. An analysis of RC compensation would require consideration of the nature of Selected Reserve RC service, where members typically are employed full-time in a civilian job or attending school, and are employed only part-time in the RC. Such an analysis is beyond the scope of the current study.

[2] In this report we do not ask the question of whether the 70th percentile is still the correct level for military pay and instead focus on correctly measuring RMC compared with civilian wages. However, in ongoing work we do explicitly explore whether the standard set by the 9th QRMC is still the appropriate benchmark.

Like the 9th QRMC, this report also focuses on active-component personnel, and it measures military pay via RMC. We compare RMC of enlisted personnel with the annual earnings of full-time, full-year workers with high school degrees and those with some college education. In our main results we adjust the civilian earnings distribution to resemble the gender mix of the military. We compare RMC for officers with college graduates and to those with advanced degrees, again adjusting the civilian distribution to look more similar to the military distribution in terms of gender.[3]

In this report we address three main questions:

- How does military pay for active-component personnel in 2017 compare with civilian pay?
- What has happened to recruit quality given the relative changes in military and civilian pay since 2000?
- How does the difference between civilian and military pay change across geographies within the United States?

In addressing these questions, we used data from the U.S. Department of Defense's (DoD's) *Selected Military Compensation Tables* (Directorate of Compensation, 2017), also known as the Greenbook, and from Active Duty Pay Files provided by the Defense Manpower Data Center (DMDC). We also use data from the March supplements to the Current Population Survey (CPS; U.S. Census Bureau, 2018), from DMDC's August 2009 and September 2017 Status of Forces Surveys on the education distribution of enlisted personnel and officers (Office of People Analytics, 2017, 2018), and from *2015 Demographics: Profile of the Military Community* (Office of the Deputy Assistant Secretary of Defense for Military Community and Family Policy, 2016) on the gender mix in the military. We weighted civilian workers by the military gender mix then computed a civilian wage distribution for each

---

[3] Note that throughout the report we focus on individual comparisons and provide no evidence on how total household military pay compares with total household civilian pay.

level of education to make results comparable with previous QRMCs.[4] Treating RMC as though it were a wage, we found its placement in the distribution (i.e., we determined its percentile). We computed RMC percentiles for officers and enlisted by year of service, as well as overall RMC percentiles for officers and enlisted for 2017 and 2009.

To examine how quality changes as civilian and military pay varies, we estimated regression models to study the relationship between recruiting outcomes and the ratio of RMC to the median civilian wage of high school completers ages 18–22, controlling for other variables. We estimated separate models by branch of service for non–prior service (NPS) accessions and used two types of recruiting outcomes, the recruiting rate and the share of accessions who are not high school diploma graduates (HSDGs). We calculated the outcomes for Armed Forces Qualification Test (AFQT) score categories I, II, IIIA, and IIIB. For instance, the category II recruiting rate in a year is the ratio of HSDG accessions in category II to the population of high school completers in that category net of those going on to complete four or more years of college. The share of non-HSDG accessions in category II is the ratio of non-HSDG accessions in category II to the total number of accessions in that category (HSDG and non-HSDG).

To examine how military wages compare with civilian wages across geographies, we build on recent work in economics that has documented fundamental changes in wage patterns between rural and urban areas.[5] Using data from the 2010 U.S. Census, we split states into the ten most urban and the ten least urban and compare RMC with civilian wages for each of these groups.

---

[4] We do not weight by race, as our results would not be directly comparable with the results of studies in support of previous QRMCs. However, military pay does not differ by race, and the military tends to be more diverse than the civilian population. Thus, if we were to weight the civilian population by the military racial mix our RMC percentiles would likely be higher than they presently are. In this way, our results are conservative estimates (that is, they are likely biased downward).

[5] See, for instance, Autor (2019).

The Regular Military Compensation Percentile for 2017 Was Above the 70th Percentile and Was About the Same as in 2009

Taking a weighted average across education levels based on the military education distribution for the first 20 years of service, we find that RMC for 2017 was at the 85th percentile of the civilian wage distribution for enlisted personnel and at the 77th percentile of the civilian wage distribution for officers.[6]

Many military members increase their educational attainment while in service, and this changes the mix of nonmilitary jobs that they can get. For this reason, it is important to compare military RMC with the pay of civilians with more years of formal education as enlisted members progress through their careers. For enlisted, RMC is above the 90th percentile during the first nine years of service (YOS), when we are comparing enlisted members with civilians with a high school degree; is around the 84th percentile in years 10–19, when we are comparing enlisted members with civilians with some college; and climbs from the 59th percentile to the 71st percentile between years 20 and 30, when we are comparing enlisted members with civilians with a bachelor's degree. RMC for officers is around the 85th percentile in years 1–9, when we are comparing officer members with civilians with a bachelor's degree, and climbs from the 69th percentile to the 77th percentile from years 10–30, when we are comparing officer members with civilians with more than a bachelor's degree.

Over time, the educational attainment of military personnel has increased, such that those in higher grades have reached higher levels of educational attainment than they did in 2009. This increase in educational attainment could potentially change the RMC percentiles

---

[6]   This is somewhat less than the 90th percentile reported by the 11th QRMC for enlisted and the 83rd percentile reported for officers. The differences come from differences in methodology as explained in Hosek et al. (2018, pp. 10–16, 28–30): namely, the method used to calculate years of experience, and weighting wages by the civilian distribution of educational attainment rather than the military distribution of educational attainment. For a discussion of the "pay gap" and how analysis of changes in basic pay and the Employment Cost Index (ECI) compare with our results, see Hosek et al. (2018, pp. 99–104).

over time. Yet we found that the overall RMC percentiles for 2017 for enlisted personnel and officers were very similar to those for 2009. This finding uses five levels of education for enlisted (high school, some college, associate's degree, bachelor's degree, and master's degree or higher), and it uses two levels for officers (bachelor's degree and master's degree or higher).

Our RMC percentile for officers—the 77th percentile in both 2009 and 2017—is below the 11th QRMC's estimate for 2009, which was the 83rd percentile. However, our methodology was also different in that we used additional education categories and imputed civilian labor force experience differently. When we compute percentiles in 2009 using a method comparable with the one we used in 2017 and include the additional education categories, we find that enlisted RMC is at around the 84th percentile in 2009, similar to our estimate for 2017. Put differently, enlisted RMC relative to civilian pay has remained unchanged between 2009 and 2017, and the differences reported here versus the 11th QRMC are attributable to differences in methodologies.

We also compared RMC with civilian wages from 2000 to 2017 for selected age and education groups. There is a steady increase in RMC relative to civilian pay from 2000 to 2010 and a leveling off afterward. This is likely due to wage stagnation in the civilian sector and continued growth in wages for military personnel through the 2000s.

Recruit Quality Rose in Three Services as Military Pay Increased Relative to Civilian Pay

Our regression findings show similar patterns to those noted in Hosek et al. (2018, pp. 52–63): namely, a positive association between enlisted recruit quality and the ratio of RMC to the civilian wage for the Navy, Marine Corps, and Air Force but not for the Army. Recruit quality is defined as individuals who enlist who are in the top half of the distribution of AFQT scores and who are HSDGs. Those who are assigned AFQT categories I, II, and IIIA are considered in the top half of the distribution. The Navy, Marine Corps, and Air Force increased quality over time as both wages and the recruiting rates for categories I and II increased. The Marine Corps also increased the recruiting rate for categories IIIA and IIIB. As wages rose, the Army decreased the recruiting rate for category IIIB, as well as for II and IIIA. The reasons for the difference in the relationship between RMC and quality for the Army are unclear. However, some potential reasons are discussed below in this report (see also Hosek et al., 2018, pp. 71–73).

Further, the Army and the Marine Corps had positive associations between the share of accessions that were non-HSDGs and the ratio of RMC to the civilian wage in categories I, II, and IIIB but not in IIIA. These services took more non-HSDGs as military pay rose, other things being equal. The Navy increased the share of non-HSDGs in categories I and II but not IIIA or IIIB. The Air Force decreased the share of non-HSDGs in category I and increased the share in categories IIIA and IIIB.

Geography Matters Less for Service Members at Lower Levels of Education and More for Service Members with Higher Levels of Education

Confirming trends noted by Autor (2019), we find that civilian wages do not differ much across geographies in the United States for workers with less than a high school degree, a high school degree, or some college, but that they differ substantially for workers with a bachelor's degree or higher. Thus, unlike in the past when civilian wages for both highly skilled and less-skilled workers were higher in urban than less-urban areas, civilian wages are more equal across geographic areas, at least for those with a high school degree or some college. We find that RMC percentiles of the civilian wage distribution for enlisted personnel with lower levels of education are similar across the most urban and the least urban states. However, RMC percentiles of the civilian wage distribution for Army officers with higher levels of education are much lower in the most urban states compared with the least urban states. As automation and outsourcing have changed the labor market and replaced many jobs that required specialized training but lower levels of formal education (such as factory jobs), alternatives for less educated workers have changed (Acemoglu and Autor, 2011; Acemoglu and Restrepo, 2017, 2018; Alabdulkareem et al., 2018; Autor, 2015, 2019; Autor and Dorn, 2013; Autor, Katz, and Kearney, 2006; Autor, Levy, and Murnane, 2003). Whereas previously many less-educated workers worked in these "middle-skill" jobs as well as less skill-intensive jobs, they are increasingly taking less skill-intensive jobs as the middle-skill jobs have disappeared. Additional research should be conducted to further examine how RMC compares with civilian wages in different parts of the country for workers with different levels of education and the implications for recruiting and retention of military personnel.

Acknowledgments

We would like to thank Thomas K. Emswiler, Director, 13th Quadrennial Review of Military Compensation, for sponsoring this study. We especially appreciate the guidance offered by Jeri Busch, Director for Military Compensation, and Don Svendsen of the Office of Compensation, as well as Don's comments. We are grateful to Mike DiNicolantonio and his team at the Research, Surveys, and Statistics Center of the Office of People Analytics in the Defense Human Resources Activity for tabulations on educational attainment of those in the military. At RAND, Christine DeMartini helped process the military pay and Current Population Survey files, and we are thankful for her assistance. We appreciate the input and comments from the two reviewers, John Warner, Professor Emeritus from Clemson University, and Melanie Zaber at RAND.

Abbreviations

AFQT	Armed Forces Qualification Test
ASVAB	Armed Services Vocational Aptitude Battery
BAH	basic allowance for housing
BAS	basic allowance for subsistence
CPS	Current Population Survey
DMDC	Defense Manpower Data Center
DoD	U.S. Department of Defense
ECI	Employment Cost Index
FY	fiscal year
GED	General Education Development
HSDG	high school diploma graduate
MEPS	military enlistment processing station
NCES	National Center for Education Statistics
NLSY	National Longitudinal Survey of Youth
NPS	non–prior service

NR	not reported
QRMC	Quadrennial Review of Military Compensation
RMC	regular military compensation
YOS	years of service

CHAPTER ONE

Introduction

A common theme of past Quadrennial Reviews of Military Compensation (QRMCs), as far back at the first one in 1969, is whether military compensation is set high enough to attract and retain the number and quality of personnel required by the armed services. Basic pay is the foundation of military compensation. Every service member on active duty is entitled to basic pay, though the particular amount depends on the member's pay grade and length of service. Every member is also entitled to receive two other elements of military compensation, the basic allowance for housing (BAH) (or quarters in kind) and basic allowance for subsistence (BAS) (or subsistence in kind). The entitlement to these three elements—basic pay, BAH, and BAS—led the Gorham Commission in 1962 to develop the construct of "regular military compensation," or RMC, as a benchmark for comparing military compensation with civilian compensation. Later, the definition of RMC was expanded to include the federal tax advantage associated with receiving BAH and BAS tax-free.

Subsequent QRMCs and commissions also considered the competitiveness, effectiveness, and efficiency of military compensation, focusing on various elements of compensation to include not only RMC but also BAH, military retirement reserve compensation, and the structure of the pay table.[1] It was the 9th QRMC that made the level of RMC the focal point of its study. In its 2002 report, the 9th

---

[1] A history of studies considering major structural changes to military compensation is provided in Appendix III of Office of the Under Secretary of Defense for Personnel and Readiness (2018).

QRMC concluded, "Pay at around the 70th percentile of comparably educated civilians has been necessary to enable the military to recruit and retain the quantity and quality of personnel it requires" (Office of the Under Secretary of Defense for Personnel and Readiness, 2002, p. xxiii). Further, it found that comparing enlisted personnel with civilians with a high school diploma no longer reflected the education level of the force because an increasing fraction of the enlisted force had some college education and the military actively recruited from the college-bound youth market. Thus, the 9th QRMC argued that comparative pay analyses should look at military pay for enlisted personnel relative to the 70th percentile of pay of civilians with some college. Similarly, the comparison group for officers should be civilians with a bachelor's degree or higher (rather than those with only a bachelor's degree).

Using data from 2009, a decade after the data used by the 9th QRMC, the 11th QRMC found that military pay exceeded the 70th percentile. Specifically, it found that RMC was at about the 90th percentile for enlisted members and at the 83rd percentile for officers. Thus, over the course of the 2000s, military pay increased substantially relative to civilian pay.

In a recent study, Hosek et al. (2018) found that military pay continued to exceed the 70th percentile and that the percentiles for 2016 were in fact virtually the same as what the 11th QRMC found for 2009. The Hosek et al. (2018) study also analyzed the extent to which readiness, as measured by the quality of the enlisted force, improved as military pay relative to civilian pay increased since 2000. The study found that recruit quality rose as relative military pay increased since 2000 for each service, except for the Army. The reason for the Army difference was unclear. Some proposed explanations include Army recruiting becoming more difficult than other services' recruiting over this period, the Army reduced other recruiting resources such as bonuses, or the Army chose to focus its recruiting efforts on nontraditional metrics of quality.

The director of the 13th QRMC requested that RAND update the Hosek et al. (2018) study to consider how military pay compares with the pay of similar civilians through 2017 rather than 2016. He also requested that we update the regression analysis in that study to examine how recruit quality changes with increases in relative military pay. In the spirit of the 9th QRMC, we first consider the educational attainment of the enlisted and officer forces and how attainment has changed since the 9th QRMC considered this question using data from 1999 and since the 11th QRMC considered this question using data from 2009. For this analysis we use input from the Office of People Analytics within the Office of the Under Secretary of Defense for Personnel and Readiness. Given the shift in the educational attainment of the enlisted and officer force, we then address three main questions:

1. How does military pay for active-component personnel in 2017 compare with the pay of comparably educated civilians? Similarly, are our results for 2017 different from the results in Hosek et al. (2018) for 2016?
2. Has the comparability of military pay changed since 2009, when the 11th QRMC compared military and civilian pay?
3. What has happened to recruit quality given the relative changes in military and civilian pay since 2000?

We address these questions using data from the U.S. Department of Defense's (DoD's) *Selected Military Compensation Tables* (Directorate of Compensation, 2017), also known as the Greenbook, and from Active Duty Pay Files provided by the Defense Manpower Data Center (DMDC). We also use data from the March supplements to the Current Population Survey (CPS; U.S. Census Bureau, 2018) and from *2015 Demographics: Profile of the Military Community* (Office of the Deputy Assistant Secretary of Defense for Military Community and Family Policy, 2016) on the gender mix in the military and from DMDC's August 2009 and September 2017 Status of Forces Surveys on the education distribution of enlisted personnel and officers (Office of People Analytics, 2017, 2018). To analyze how recruit quality changes as civilian and military pay varies, we estimated regression models for each service, controlling for other variables that change over time (such as unemployment rate and recruiting goals).

Chapter Two shows how educational attainment for military personnel has changed over time and then addresses the first two questions above. We compare RMC with civilian wages both over a career and over calendar time for specific age and education groups. Chapter Three summarizes our analysis for the third question regarding the relationship between recruit quality and relative military pay. Because our methodology for both Chapters Two and Three closely follows the methodology used in Hosek et al. (2018), we provide only a broad overview of our methods and focus more on results. Interested readers are referred to the Hosek et al. (2018) document. In Chapter Four, we explore how geography affects the competitiveness of military pay and how this varies by education level. We offer concluding thoughts in Chapter Five.

CHAPTER TWO

Comparisons of Military and Civilian Pay

In this chapter we examine how RMC compares with civilian pay. RMC includes basic pay, BAH, BAS, and the federal tax advantage resulting from the allowances not being taxed. RMC accounts for approximately 90 percent of current cash compensation (Office of the Under Secretary of Defense for Personnel and Readiness, 2012a, 2012b).

We note that throughout the report we are comparing the pay of individuals and not households and that we will not provide any evidence on the adequacy of military pay for military families. Previous work has documented that military spouses have significantly lower rates of employment and earnings than comparable civilians and that they tend to be underemployed when employed (Asch, Hosek, and Warner, 2007). Spousal employment and earnings have increased as a share of family employment and earnings over time, and this may affect military and civilian families differently. We are conducting ongoing work that will more fully examine the adequacy of military pay.

We first examine how educational attainment for military personnel has changed over time and use these data to adjust our measures of RMC. We compare RMC with civilian wages both over a career and through time for specific age and education groups.

Educational Attainment

To compare military personnel with similar civilians, we used the education distribution of officers and enlisted personnel from the August 2009 and September 2017 Status of Forces Surveys of Active

Duty Members, provided by the DoD Office of People Analytics.[1] We considered five education levels for enlisted personnel—high school, some college (more than high school but no degree), associate's degree, bachelor's degree, and master's degree or higher—and two levels for officers—bachelor's degree and master's degree or higher.

In Table 2.1 we show the education attainment for enlisted members in 2017 and in 2009 when the 11th QRMC compared military and civilian pay. We find that enlisted personnel in 2017 and 2009 show a similar education profile, although those in 2017 have more years of formal education than those in 2009. The 9th QRMC identified the trend of increasing educational attainment while members were in service, and we find evidence that this trend continued beyond 1999.

Table 2.1
Educational Attainment of Enlisted Personnel, by Pay Grade, 2009 and 2017, as Percentages

Pay Grade	Non–High School Graduate		High School Graduate		Less Than One Year of College		One or More Years of College, No Degree		Associate Degree		Bachelor's Degree		Advanced Degree	
	2009	2017	2009	2017	2009	2017	2009	2017	2009	2017	2009	2017	2009	2017
E-2	1	1	70	66	20	20	8	13	NR	0	1	0	NR	0
E-3	1	1	48	51	23	16	21	22	4	5	3	4	0	0
E-4	0	0	39	40	25	18	22	23	7	9	6	8	1	1
E-5	1	1	25	23	22	18	32	30	13	18	6	9	0	1
E-6	1	0	17	13	23	15	30	32	20	26	8	12	1	2
E-7	1	0	10	9	15	10	30	22	28	32	14	22	2	5
E-8	0	0	9	4	13	7	30	25	24	24	20	28	4	11
E-9	NR	0	7	6	10	4	17	14	22	22	30	37	14	18

SOURCES: Office of the Under Secretary of Defense for Personnel and Readiness, 2002, Figure 2.5; Office of People Analytics, 2017, 2018.

NOTE: NR = not reported. The percentages in each row add to 100 with rounding. There is no row for E-1s because their education distribution was not reported in the survey. In this table, *high school graduate* includes traditional diploma and alternative diploma (e.g., home school, equivalency test, distance learning). The survey responses are weighted to be representative of the force.

---

[1]   These data were provided to RAND by the Office of People Analytics in 2017 and 2018, respectively.

In Table 2.2 we show how the educational attainment of enlisted personnel has changed from 1999 to 2017. Unfortunately, we do not have detailed information on all education categories for 1999, and so we present tabulations for the two groups we do have: those with some college or an associate's degree and those with a bachelor's degree or higher. The year 1999 is when the 9th QRMC compared military and civilian pay. In 1999 18 percent of E-2s had some college or higher education; for E-9s the figure was 84 percent. For 2009, when the 11th QRMC compared military and civilian pay, the percentages were 29 percent and 93 percent, respectively, and for 2017 they were 33 percent and 95 percent respectively. Many military members increase their educational attainment while in service, and this changes the mix of nonmilitary jobs that they can get. For this reason, it is important to

Table 2.2
Enlisted Personnel with Post–High School Education, by Pay Grade, 1999, 2009, and 2017, as Percentages

Pay Grade	Some College or Associate's Degree			Bachelor's Degree or Higher		
	1999	2009	2017	1999	2009	2017
E-1	7	NR	NR	1	NR	NR
E-2	18	28	33	0	1	0
E-3	22	48	43	2	3	4
E-4	31	54	50	5	7	9
E-5	47	67	66	6	6	10
E-6	57	73	73	10	9	14
E-7	60	73	64	18	16	27
E-8	56	67	56	22	24	39
E-9	57	49	40	27	44	55

SOURCES: Office of the Under Secretary of Defense for Personnel and Readiness, 2002, Figure 2.4; Office of People Analytics, 2017, 2018.

NOTE: NR = not reported. There is no data for E-1s after 1999 because their education distribution was not reported in the survey. The survey responses are weighted to be representative of the force. The 9th QRMC report presents the combined percentage of enlisted with bachelor's degrees or higher; it does not present the percentage with only a bachelor's degree. For 2009 and 2017, Table 2.1 shows separate percentages for bachelor's degrees and master's degrees or higher, and this table adds those percentages to obtain bachelor's degrees or higher.

Table 2.3
Educational Attainment of Officer Personnel, by Pay Grade, 1999, 2009, and 2017, as Percentages

Pay Grade	College Degree			Advanced Degree		
	1999	2009	2017	1999	2009	2017
O-1	97	93	91	3	6	8
O-2	91	87	87	9	11	12
O-3	59	60	57	39	39	42
O-4	31	30	20	69	69	79
O-5	15	13	7	85	85	93
O-6	8	4	2	92	96	98

SOURCES: Office of the Under Secretary of Defense for Personnel and Readiness, 2002, Figure 2.14; Office of People Analytics, 2017, 2018.

NOTE: *College degree* includes bachelor's and associate's degrees. *Advanced degree* includes master's, doctoral, and professional school degrees.

compare military RMC with the pay of civilians with more years of formal education as enlisted progress through their careers.

As with their enlisted counterparts, the percentage of officers with bachelor's degree or higher increases with rank and has trended upward over time. Table 2.3 shows the percentage of officers with college degrees and advanced degrees for 1999, 2009, and 2017. The percentage of O-1s with an advanced degree increased from 3 percent in 1999 to 8 percent in 2017, and the percentage of O-6 with an advanced degree increased from 92 percent to 98 percent over the time period.

Regular Military Compensation Percentiles in 2017

In this section we consider how RMC compares with the pay of similar civilians over a career and overall, averaged across all personnel. For this analysis, we use RMC from the Directorate of Compensation's *Selected Military Compensation Tables*, or Greenbook (Directorate of Compensation, 2017). In it, RMC is an average across pay grade and dependency status at each year of service. Data on weekly wages and characteristics for civilians come from the Current Population Survey

(CPS) Annual Social and Economic Supplement, also known as the March CPS. The CPS, administered by the Bureau of Labor Statistics, uses a representative random sample of the population.

Following the 11th QRMC, we used data on full-time, full-year workers and weight civilian-wage data by the percentages of men and women in the military.[2] In 2015, the percentages were 85 percent men and 15 percent women for enlisted and 83 percent men and 17 percent women for officers (Office of the Deputy Assistant Secretary of Defense for Military Community and Family Policy, 2016, pp. 18–19).

While the Greenbook provides RMC by years of service (YOS), the CPS does not have data on civilian years of labor-force experience. To compare military and civilian wages adjusted for experience, we used assumptions to map age and years of education to years of labor-force experience for civilians. Specifically, for high school graduates we subtracted 18 from the person's age in years, for those with some college and associate's degrees we subtracted 20, for college graduates we subtracted 22, and for those with advanced degrees we subtracted 24. For those who started school at a later age or who interrupted their schooling for any reason, these assumptions overstate their experience.[3] However, most students initially enrolling in two- and four-year institutions are 19 years or younger (National Center for Education Statistics [NCES], 2017).[4]

Regular Military Compensation Percentiles over a Career in 2017

Figures 2.1a, 2.1b, and 2.2 show enlisted and officer RMC in 2017 for a given year of service and compare RMC with civilian wages for a comparable year of experience at the 50th (median) and 70th percentiles, for the levels of education noted in the figure notes. As shown in Table

---

[2]   These are workers with a usual workweek of more than 35 hours and who worked more than 35 weeks in the year.

[3]   Since we are treating "some college" as two years, we may also be underestimating work experience for some individuals. However, for those who start school late, take a gap year, complete extended religious mission service, or take more than four years for college or more than two years for graduate school, we are assigning them more experience than they have.

[4]   For more information on how these assumptions impact our estimates as well as details about top coding in the CPS and how this approach compares with that of the 11th QRMC, see Hosek et al. (2018, p 17).

Figure 2.1a
Enlisted Regular Military Compensation, Civilian Wages, and Regular Military Compensation Percentiles for Full-Time, Full-Year Workers with High School, Some College, or Bachelor's Degree, 2017

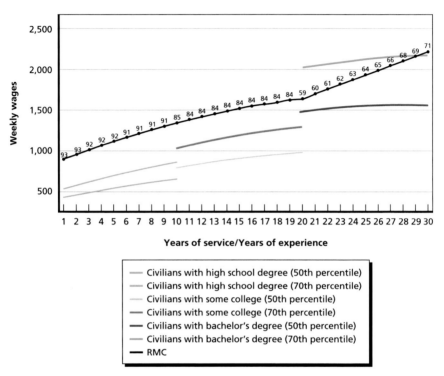

SOURCES: Directorate of Compensation, 2017; U.S. Census Bureau, 2018.
NOTE: RMC percentile varies by YOS (1–9 = high school, 10–19 = some college, and 20–30 = a bachelor's degree). We weighted civilian-wage data by enlisted military gender mix. Colored lines are smoothed wage curves for the 50th and 70th percentiles of the given level of education. The black line is enlisted RMC, and the number above the black line is the percentile in the wage distribution for high school, some college, and bachelor's degree.

2.2, 39 percent of E8s and 55 percent of E9s had a bachelor's degree or higher in 2017. In Figure 2.1a we compare enlisted RMC for YOS 20–30 to civilian wages for civilians who have a bachelor's degree or higher. In Figure 2.1b we compare enlisted RMC for YOS 20–30 to civilian wages for civilians who have an associate's degree. In Table 2.4 we estimate the education distribution for each YOS to provide a more precise picture of overall RMC. The civilian-wage and RMC lines in Figures 2.1a, 2.1b, and 2.2 have been smoothed with quadratic regressions on the raw data.

Figure 2.1b
Enlisted Regular Military Compensation, Civilian Wages, and Regular Military Compensation Percentiles for Full-Time, Full-Year Workers with High School, Some College, or Associate's Degree, 2017

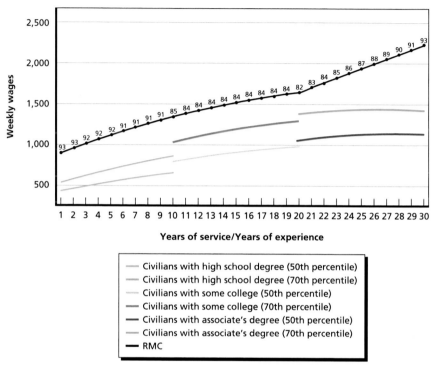

SOURCES: Directorate of Compensation, 2017; U.S. Census Bureau, 2018.
NOTE: RMC percentile varies by YOS (1–9 = high school, 10–19 = some college, and 20–30 = a associate's degree). We weighted civilian-wage data by enlisted military gender mix. Colored lines are smoothed wage curves for the 50th and 70th percentiles of the given level of education. The black line is enlisted RMC, and the number above the black line is the percentile in the wage distribution for high school, some college, and associate's degree.

Compared with civilians with a high school degree, enlisted pay is around the 90th percentile of the civilian pay distribution in the first part of the career (1–9 YOS). When we compare RMC with civilians with some college for years 10–19, enlisted pay is at around the 84th percentile. RMC then rises from the 59th percentile to the 71st percentile for YOS 20–30 when compared with the pay of civilians with a bachelor's degree (Figure 2.1a). When we compare to civilians with an

Figure 2.2
Officer Regular Military Compensation, Civilian Wages, and Regular Military Compensation Percentiles for Full-Time, Full-Year Workers with Bachelor's Degree or with Master's Degree or Higher, 2017

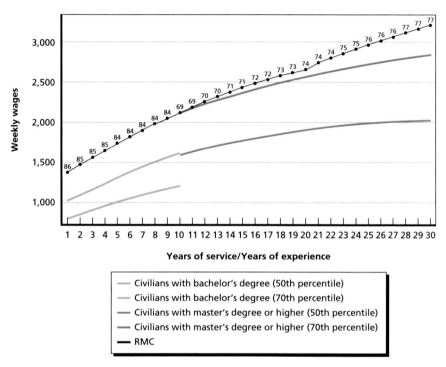

SOURCES: Directorate of Compensation, 2017; U.S. Census Bureau, 2018.
NOTE: RMC percentile varies by YOS (1–9 = bachelor's degree, 10–30 = master's degree or higher). We weighted civilian-wage data by officer gender mix. Colored lines are smoothed wage curves for the 50th and 70th percentiles of the given level of education. The black line is enlisted RMC, and the numbers above the black line are the percentile in the wage distribution for a bachelor's degree and a master's degree or higher.

associate's degree (Figure 2.1b), pay goes from the 82nd percentile in YOS 20 to the 93rd percentile in YOS 30. It is not surprising that the percentiles are lower for 20–30 YOS than for the YOS earlier in the career because personnel policies become more selective in terms of which personnel are allowed to be retained after 20 YOS—that is, the point at which military personnel are eligible for an immediate annuity under the military retirement system. As shown in the figures, RMC sharply increases after 20 YOS, reflecting the higher quality and higher pay of those permitted to stay on. For 20–30 YOS in Figure 2.1a, we are comparing RMC with the wages of civilians with a bachelor's degree, and this results in lower RMC percentiles of the civilian wage distribution. When compared to civilians with an associate's degree, RMC percentiles remain relatively high through YOS 30.

Officer RMC is at about the 85th percentile of the civilian wage distribution when compared with civilians with a bachelor's degree in the early career (years 1–9) and rises from the 69th percentile to the 77th percentile from years 10–30 when compared with civilians with a master's degree or higher.

Weighted Average of Regular Military Compensation Percentiles for 2017

It is also of interest to have an overall summary measure of how RMC compares with civilian pay, so we computed an overall weighted average of the RMC percentiles. The procedure for estimating the overall average is provided in Hosek et al. (2018, pp. 26–28), but in short, we computed the RMC percentiles for 2017 by YOS for each education level. This way, we could examine the RMC percentile in detail by education level. In addition, we estimated the percentage of education distribution at each year of service, used this to compute the average RMC percentile by YOS, and then used the number of personnel by YOS to compute an overall weighted average of the RMC percentile.[5] To compute percentiles, civilian pay by formal education level and age

---

[5] To compute the weighted averages, we first translated the education distribution by rank (Tables 2.1 and 2.3) to a distribution at each year of service, by interpolation. We did this in several steps. First, we obtained the joint distribution of personnel by pay grade and YOS from the Greenbook. This allowed us to compute the percentage of personnel at each pay grade, by YOS. Second, we used these percentages to obtain a weighted average of the education distribution at each year of service (i.e., the percentage with high school, some college, associate's degrees, bachelor's degrees, and master's degrees or higher). Third, for each level of education (e.g., high school, some college), we fitted a polynomial curve to its percentages by YOS and then used the fitted curves to predict the percentage, in effect smoothing the percentages. The set of curves for the different levels of education gave us the predicted education distribution by YOS. The predicted education distribution is shown in Tables 2.4 and 2.5 for enlisted and officers, respectively. To check for sensitivity, we perturbed the education percentages by YOS and found little change in the predicted overall RMC percentile.

Table 2.4
Regular Military Compensation as a Percentile of Civilian Wages, by Level of Education and Year of Service, for Enlisted Personnel, 2017

YOS	Predicted Education Distribution, by YOS					RMC Percentile					Weighted Average	Enlisted Count
	High School	Some College	Associate's	Bachelor's	Master's Plus	High School	Some College	Associate's	Bachelor's	Master's Plus		
1	0.56	0.37	0.04	0.04	0.00	92	88	79	65	32	88.9	175,468
2	0.46	0.40	0.08	0.06	0.00	92	90	84	53	33	88.1	133,761
3	0.38	0.43	0.11	0.07	0.01	92	91	84	59	38	87.9	124,916
4	0.32	0.45	0.14	0.08	0.01	88	95	85	57	38	87.7	124,147
5	0.27	0.46	0.16	0.09	0.01	95	83	88	61	37	84.5	103,467
6	0.24	0.47	0.18	0.10	0.01	92	91	81	61	49	85.9	71,618
7	0.21	0.47	0.20	0.10	0.02	93	88	87	57	39	84.9	56,219
8	0.19	0.47	0.21	0.11	0.02	89	91	77	63	32	83.6	46,891
9	0.17	0.46	0.23	0.12	0.02	92	88	77	55	39	81.5	39,891
10	0.16	0.46	0.24	0.12	0.02	90	81	83	55	40	78.9	33,298
11	0.15	0.45	0.25	0.13	0.02	92	82	78	60	35	78.6	28,268
12	0.14	0.44	0.26	0.14	0.02	93	83	82	57	34	79.3	28,359
13	0.13	0.42	0.27	0.15	0.03	86	88	79	56	37	79.1	22,956
14	0.12	0.41	0.28	0.16	0.03	92	82	82	61	41	78.6	24,819
15	0.11	0.40	0.28	0.17	0.04	94	79	74	52	32	73.0	23,884
16	0.11	0.38	0.29	0.18	0.04	91	80	85	61	36	77.3	22,299
17	0.10	0.37	0.29	0.19	0.05	89	84	83	55	41	76.5	21,128
18	0.09	0.36	0.29	0.21	0.06	91	81	79	55	38	73.6	20,980

Table 2.4—Continued

| YOS | Predicted Education Distribution, by YOS | | | | | RMC Percentile | | | | | | |
---	High School	Some College	Associate's	Bachelor's	Master's Plus	High School	Some College	Associate's	Bachelor's	Master's Plus	Weighted Average	Enlisted Count
19	0.08	0.35	0.28	0.22	0.06	91	82	81	51	38	72.9	19,181
20	0.08	0.34	0.28	0.23	0.07	86	87	87	65	41	78.4	19,093
21	0.07	0.32	0.27	0.24	0.08	89	87	79	55	42	73.4	19,052
22	0.07	0.31	0.27	0.26	0.09	92	87	80	60	43	74.4	10,559
23	0.07	0.30	0.26	0.27	0.10	93	86	83	60	47	74.6	7,756
24	0.07	0.28	0.25	0.29	0.11	94	87	86	61	47	75.2	6,492
25	0.06	0.27	0.25	0.30	0.12	92	84	81	62	48	72.7	5,412
26	0.06	0.25	0.24	0.31	0.13	92	91	92	68	55	79.3	3,943
27	0.06	0.24	0.24	0.32	0.14	96	90	95	69	48	78.8	2,730
28	0.06	0.22	0.23	0.33	0.15	97	89	89	70	58	78.6	2,117
29	0.06	0.22	0.23	0.33	0.15	97	91	91	71	55	79.2	1,971
30	0.06	0.22	0.24	0.33	0.15	95	90	92	69	54	78.3	1,471
0–20											84.6	
0–30											84.1	

SOURCES: Directorate of Compensation, 2017; DMDC data from 2017; U.S. Census Bureau, 2018.

NOTE: We computed the RMC percentile at each level of education, by YOS, as median RMC relative to the civilian wages of full-time, full-year male and female workers, weighted by their proportion in the military. We computed median RMC from the Greenbooks with weights based on the fraction of personnel count, by YOS (the "Enlisted Count" column), which comes from the active-duty pay files. Weighted average RMC percentile at each year of service is the sum of the product of the RMC percentile at a given level of education and the fraction of personnel with that level of education, shown in the left pane of the table. We estimated the education fractions using the educational attainment distribution for 2017 (see Table 2.1) and the joint distribution of personnel by pay grade and YOS from the Greenbook for 2017. The overall RMC percentiles for YOS 0–20 and YOS 0–30 are weighted averages of the average RMC percentile at each year of service, with weights based on the fraction of personnel count by YOS (the "Enlisted Count" column).

Table 2.5
Regular Military Compensation as a Percentile of Civilian Wages, by Level
of Education and Year of Service, for Officers, 2017

	Predicted Education Distribution[a]		RMC Percentile			
YOS	Bachelor's	Master's Plus	Bachelor's	Master's Plus	Weighted Average	Officer Count
1	0.92	0.08	81	55	78.9	10,242
2	0.84	0.16	85	51	79.5	9,299
3	0.76	0.24	86	68	81.7	9,244
4	0.69	0.31	87	74	83.0	9,132
5	0.63	0.37	86	72	80.8	9,829
6	0.56	0.44	82	73	78.1	9,077
7	0.51	0.49	85	68	76.6	8,794
8	0.45	0.55	88	64	74.9	8,464
9	0.41	0.59	81	70	74.5	8,034
10	0.36	0.64	85	77	79.9	7,842
11	0.32	0.68	85	75	78.2	7,569
12	0.28	0.72	81	67	71.0	6,909
13	0.25	0.75	86	75	77.7	6,302
14	0.22	0.78	84	74	76.2	6,556
15	0.19	0.81	82	68	70.7	6,340
16	0.17	0.83	87	67	70.4	6,280
17	0.15	0.85	84	70	72.1	6,252
18	0.13	0.87	83	73	74.3	6,261
19	0.11	0.89	78	71	71.8	5,882
20	0.10	0.90	89	71	72.8	6,088
21	0.09	0.91	85	65	66.8	6,138
22	0.08	0.92	83	74	74.7	4,657
23	0.08	0.92	85	75	75.8	4,316
24	0.07	0.93	87	77	77.7	4,054
25	0.07	0.93	89	78	78.8	3,810
26	0.07	0.93	87	74	74.9	3,526
27	0.07	0.93	88	71	72.2	2,959
28	0.07	0.93	87	78	78.7	2,606

Table 2.5—Continued

| YOS | Predicted Education Distribution[a] | | RMC Percentile | | | Officer Count |
	Bachelor's	Master's Plus	Bachelor's	Master's Plus	Weighted Average	
29	0.08	0.92	89	81	81.6	2,507
30	0.08	0.92	90	82	82.7	1,996
0–20					**76.7**	
0–30					**76.4**	

SOURCES: Directorate of Compensation, 2017; DMDC data from 2017; U.S. Census Bureau, 2018.

NOTE: We computed the RMC percentile at each level of education, by YOS, as median RMC relative to the civilian wages of full-time, full-year male and female workers, weighted by their proportion in the military. We computed median RMC from the Greenbooks with weights based on the fraction of personnel count, by YOS (the "Officer Count" column), which comes from the active-duty pay files. Weighted average RMC percentile at each year of service is the sum of the product of the RMC percentile at a given level of education and the fraction of personnel with that level of education, shown in the left pane of the table. We estimated the education fractions using the educational attainment distribution for 2017 (see Table 2.1) and the joint distribution of personnel by pay grade and YOS from the Greenbook for 2017. The overall RMC percentiles for 0–20 YOS and 0–30 YOS are weighted averages of the average RMC percentile at each year of service, with weights based on the fraction of personnel count by YOS (the "Officer Count" column).

[a] This is the fraction of officers, by education level, at each year of service.

was drawn from the CPS, and military pay for each year of service was drawn from the Greenbooks. Table 2.4 presents the results for enlisted, and Table 2.5 presents the results for officers.

For enlisted personnel we estimate RMC to be at the 85th percentile of civilian wages for 0–20 YOS and at the 84th percentile of civilian wages for 0–30 YOS. The 11th QRMC reported their results averaged over 0–20 YOS, so we show results for both 0–20 and 0–30 YOS. Both calculations show that RMC is well above the 70th percentile of civilian pay even when accounting for the higher educational attainment of enlisted personnel since 1999.

For officers we estimate RMC to be at the 77th percentile of civilian wages when examining 0–20 YOS and at the 76th percentile of civilian wages for 0–30 YOS. Officers start their careers around the 79th percentile before dropping to around the 70th percentile around year 20 before climbing back to around the 80th percentile near year 30.[6] Again, the results show that RMC for officers exceeds the 70th percentile.

How do these results compare with those found by the 11th QRMC for 2009? That is, did military pay worsen or improve relative to civilian pay between 2009 and 2017?

Our weighted average estimates, which place RMC at the 85th percentile for enlisted and the 77th percentile for officers, are lower than the estimates of the 11th QRMC, which placed RMC at the 90th percentile of civilian pay for enlisted and the 83rd for officers. However, our methodology was also different, as we used additional education categories and computed civilian years of experience differently.[7] The issue is that we have a measure of years of service for military personnel, but the CPS does not provide a comparable measure of years of labor-force experience. So, assumptions using data on age are required to impute years of experience in the CPS data. Our assumptions differ from that of the 11th QRMC, which used a conservative estimate, with the result that it potentially missed wage growth in civilian pay, thereby causing percentiles to appear larger. Our approach is less conservative, resulting in somewhat lower percentiles. Neither approach is perfect; both yield qualitatively similar results. When we computed percentiles in 2009 using a method comparable with the one we used in 2017 and included the additional education categories, we found that enlisted RMC is at around the 84th percentile in 2009, which is similar to our estimate for 2017. Put differently, enlisted RMC relative to civilian pay has remained unchanged between 2009 and 2017, and the differences reported here versus the 11th QRMC are attributable to differences in methodology. For officers we find that the weighted average of the RMC percentile for 2009 is the 78th for years 0–20. Again, this is virtually the same as the 77th percentile we find for 2017 using the same methodology and the same education categories.

---

[6]   As mentioned, these figures are weighted averages that reflect the distribution of educational attainment among officers. The percentages shown by YOS in Figure 2.2 assume a given level of education at different YOS. Consequently, we would not expect the percentiles at any given YOS in Figure 2.2 to equal the weighted average across YOS.

[7]   For an in-depth discussion of the differences in the two approaches, see Hosek et al. (2018, pp. 10–16, 28–30).

In summary, we find little change in average RMC percentiles between 2009 and 2017 when calculated using a consistent methodology.

Trends in the Regular Military Compensation Percentile for Selected Age and Education Groups, 2000–2017

To see whether and to what extent RMC percentiles evolved over time rather than a point in time, we also computed the RMC percentile for 2000 through 2017 for specific groups defined by education level and age. We conducted this analysis for each service, but we present only the results for Army men because results were similar across services. In these graphs we use cross-section data on males from the given age group and rank (officer or enlisted) from the Defense Manpower Data Center Active Duty Pay Files to compute RMC.[8] Figures 2.3 through 2.6 show results for Army men in the following groups:

- enlisted members ages 23–27 compared to civilian high school graduates
- enlisted members ages 28–32 compared to civilians with some college
- officers ages 28–32 compared to civilians with bachelor's degrees
- officers ages 33–37 compared to civilians with master's degrees or higher.

Unlike in Figures 2.1 and 2.2, we did not smooth the wage percentiles in Figures 2.3 through 2.6. These comparisons also differ from the YOS comparisons in the earlier figures because some individuals enter service at older ages and have fewer years of service than one would expect based on their ages.

Overall, we find that RMC for these groups increased from 2000 to 2010 and then stayed roughly constant through 2017. The increase

---

[8]   Computing RMC with the military-pay files required that we compute the tax advantage. It is based on taxable (basic-pay) and nontaxable (BAS and BAH) income, number of dependents, and marital status. Additional details can be found in Hosek et al (2018, pp. 4–8).

Figure 2.3
Civilian Wages for High School Graduate Men and Median Regular Military Compensation for Army Enlisted, Ages 23–27, Calendar Years 2000–2017, in 2017 Dollars

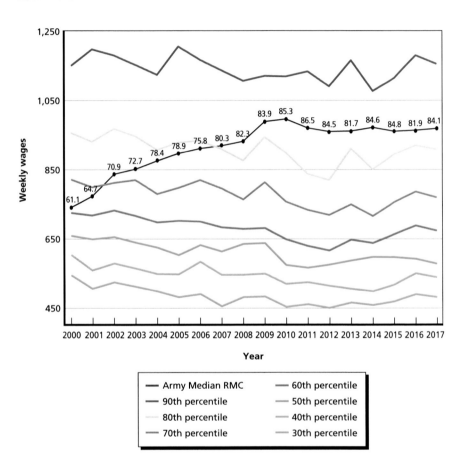

SOURCES: Active-duty pay files from DMDC; U.S. Census Bureau, 2018.
NOTE: The reference population is men ages 23–27 who reported high school completion as their highest level of education, worked more than 35 weeks in the year, and usually worked more than 35 hours per week. We computed the weekly wage by dividing annual earnings by annual weeks worked. The colored lines depict the wages at the indicated percentiles of the wage distribution for this population. For instance, at the 70th percentile, 30 percent of the population had higher wages and 70 percent had lower wages. The black line depicts median RMC for Army enlisted between ages 23 and 27. The numbers above the RMC line are the percentiles at which RMC stood in the population's wage distribution.

Figure 2.4
Civilian Wages for Men with Some College and Median Regular Military Compensation for Army Enlisted, Ages 28–32, Calendar Years 2000–2017, in 2017 Dollars

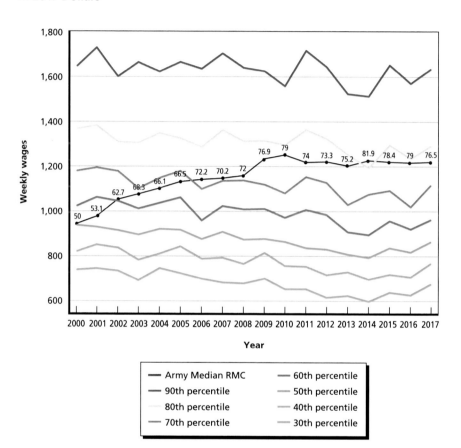

SOURCES: Active-duty pay files from DMDC; U.S. Census Bureau, 2018.
NOTE: The reference population is men ages 28–32 who reported some college as their highest level of education, worked more than 35 weeks in the year, and usually worked more than 35 hours per week. We computed the weekly wage by dividing annual earnings by annual weeks worked. The colored lines depict the wages at the indicated percentiles of the wage distribution for this population. For instance, at the 70th percentile, 30 percent of the population had higher wages and 70 percent had lower wages. The black line depicts median RMC for Army officers between ages 28 and 32. The numbers above the RMC line are the percentiles at which RMC stood in the population's wage distribution.

Figure 2.5
Civilian Wages for Men with Four-Year College Degrees and Median Regular Military Compensation for Army Officers, Ages 28–32, Calendar Years 2000–2017, in 2017 Dollars

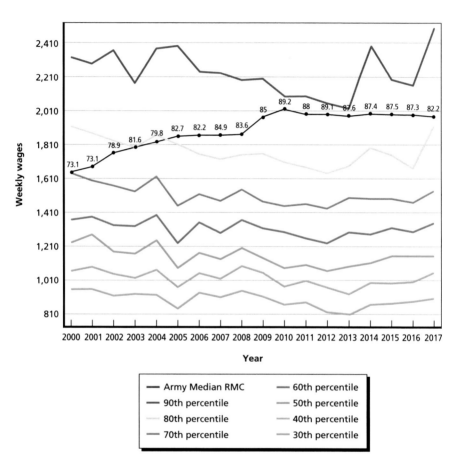

SOURCES: Active-duty pay files from DMDC; U.S. Census Bureau, 2018.
NOTE: The reference population is men ages 28–32 who reported bachelor's degrees as their highest levels of education, worked more than 35 weeks in the year, and usually worked more than 35 hours per week. We computed the weekly wage by dividing annual earnings by annual weeks worked. The colored lines depict the wages at the indicated percentiles of the wage distribution for this population. For instance, at the 70th percentile, 30 percent of the population had higher wages and 70 percent had lower wages. The black line depicts median RMC for Army officers between ages 28 and 32. The numbers above the RMC line are the percentiles at which RMC stood in the population's wage distribution.

**Figure 2.6
Civilian Wages for Men with Master's Degrees or Higher and Median
Regular Military Compensation for Army Officers, Ages 33–37, Calendar
Years 2000–2017, in 2017 Dollars**

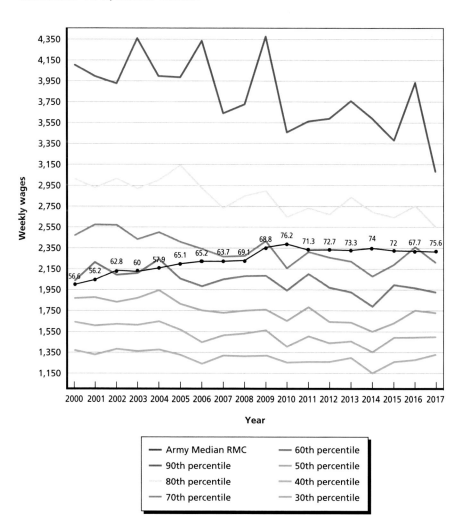

SOURCES: Active-duty pay files from DMDC; U.S. Census Bureau, 2018.
NOTE: The reference population is men ages 33–37 who reported master's degrees or higher as
their highest levels of education, worked more than 35 weeks in the year, and usually worked more
than 35 hours per week. We computed the weekly wage by dividing annual earnings by annual
weeks worked. The colored lines depict the wages at the indicated percentiles of the wage
distribution for this population. For instance, at the 70th percentile, 30 percent of the population
had higher wages, and 70 percent had lower wages. The black line depicts median RMC for Army
officers between ages 33 and 37. The numbers above the RMC line are the percentiles at which
RMC stood in the population's wage distribution.

was driven by several factors including a restructuring of the basic-pay table from 2001 through 2003, higher-than-usual basic-pay increases from (fiscal year) FY 2000 to FY 2010, increases in BAH implemented in the first part of the decade to cover the full cost of housing, increases in housing cost that resulted in further BAH increases, and a downward trend in civilian wages that leveled off around 2012 and tended to increase after 2013.[9]

From 2010 onward, the figures indicate that RMC was between

- the 81st and 87th percentiles for enlisted members ages 23–27 compared to civilian high school graduates
- the 73rd and 82nd percentiles for enlisted members ages 28–32 compared to civilians with some college
- the 82nd and 89th percentiles for officers ages 28–32 compared to civilians who were four-year college graduates
- the 68th and 76th percentiles for officers ages 33–37 compared to civilians with master's degrees or higher.

Summary

We computed RMC percentiles for enlisted and officers adjusting for the education distribution of military personnel. We estimated RMC for enlisted to be at the 85th percentile of the civilian wage distribution and RMC for officers to be at the 77th percentile of the civilian wage distribution. We also computed RMC percentiles for 2009. Our RMC percentiles are similar but somewhat lower than those of the 11th QRMC, and we attribute this to methodological differences. That is, accounting for these differences, we find little change between RMC percentile estimates for both enlisted and officers between 2009 and 2017. Trend analysis shows an increase in RMC percentile from 2000 to 2017 for various age and education groups. This reflects the relatively fast military pay growth from 2000 to 2010, as well as a downward trend in real civilian wages.

---

[9]   For more details on each of these factors, see Hosek et al. (2018, pp. 30–35).

CHAPTER THREE

Recruit Quality and Military and Civilian Pay

Military compensation is one of the primary tools used by the services to get the quantity and quality of personnel they need. Both RMC and the RMC percentile have increased substantially since 1999. Consequently, it is natural to wonder if this has resulted in an increase in the quality of recruits over time. That is, did readiness as measured by the quality of enlisted recruits increase as relative pay increased?

Military recruits are deemed high quality if they are high school degree graduates (HSDGs) and score in the upper half of the Armed Forces Qualification Test (AFQT) score distribution.[1] AFQT scores are normed to the general population using the distribution of AFQT scores from representative Bureau of Labor Statistics surveys so that they range from 0 to 99 and are subdivided into categories:

- Category I: 93–99
- Category II: 65–92
- Category IIIA: 50–64
- Category: IIIB: 31–49
- Category: IV: 16–30
- Category V: 0–15.

Thus, a recruit is in the upper half of the AFQT score distribution if he or she is in categories I–IIIA.

To examine the relationship between RMC and recruit quality, we estimated reduced-form regression models of two recruiting outcomes:

---

[1] AFQT is comprised of four sections from the Armed Services Vocational Aptitude Battery (ASVAB), which all enlisted take.

the recruiting rate for HSDGs and the non-HSDG share of accessions. We defined the outcomes separately for each of the AFQT score categories I–IIIB. We also only consider those recruits who have no prior service, that is, non–prior service (NPS) accessions.

We construct the recruiting rate as the ratio of HSDG accessions in a given AFQT category to the population of youth who have completed high school in that AFQT category, net of those who went on to complete four or more years of college. We use DMDC's Military Entrance Processing Command data for information on HSDG accession in a given AFQT category. The population of youth high school completers within a given AFQT category is estimated using a methodology described in Appendix B of Hosek et al. (2018). The methodology involves using National Center of Education Statistics (NCES) information on the population of high school completers and adjusting for AFQT category using data from the 1997 National Longitudinal Survey of Youth. Recruiting rates by service, category, and gender can be found in Appendix B Tables B.2–B.5.

The second outcome we consider is the share of non-HSDG accessions, which is computed as the ratio of non-HSDG accessions in a given AFQT category to the total number of accessions in that category (HSDG and non-HSDG). We compute this outcome using DMDC's Military Entrance Processing Command data.

We use data from 2000 through 2017. The most recent March CPS available for this analysis is for March 2018, which reports earnings for 2017, so the last year we include is 2017. Before presenting our regression results, we first show trends in recruit quality and the explanatory variables or factors that we posit are related to recruit quality and included in the regression models. We then discuss the models and results.

Trends in Recruit Quality and Factors Related to Recruit Quality

NPS recruit quality changed between 2000 and 2018. Figure 3.1 shows the percentage of accessions who are high quality, by service, defined here as HSDGs in AFQT categories I through IIIA. Figure 3.2 shows the percentages who are HSDGs, while Figure 3.3 shows the percentages who are in AFQT categories I through IIIA.

Recruit quality increased between 2000 and 2017 for the Air Force, Navy, and Marine Corps, but not the Army. The Air Force, Navy, and Marine Corps increased their percentages of accessions who were high quality (Figure 3.1) and had, or reached, a very high percentage of accessions who were NPS HSDGs (Figure 3.2). The Army's percentage of accessions who were high quality fell after 2004, then rebounded to its initial level by 2010, and then stayed there. Its HSDG percentage bottomed out in 2007 and then rose to a stable level closer that of the other services by 2010. Its percentage of accessions in categories I through IIIA in the active component declined fairly steadily after 2004 (Figure 3.2) but showed an uptick in 2018. This percentage

Figure 3.1
Percentage of Active-Component Non–Prior Service Accessions Who Were Category I–IIIA High School Diploma Graduates, by Service, FY 2000–2018

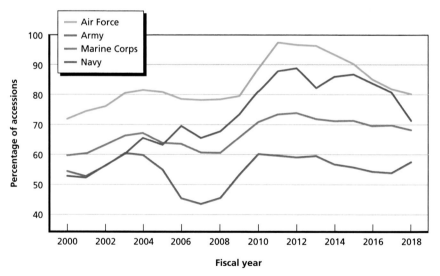

SOURCE: Office of People Analytics, undated.
NOTE: An HSDG is someone with at least a high school diploma and not exclusively a GED, associate's degree, professional nursing diploma, bachelor's degree, master's degree, post–master's degree, first professional degree, doctoral degree, post-doctorate work, or one semester of college completed. Category I–IIIA personnel are those who scored in the upper half of the AFQT score distribution.

Figure 3.2
**Percentage of Active-Component Non–Prior Service Accessions Who Were
High School Diploma Graduates, by Service, FY 2000–2018**

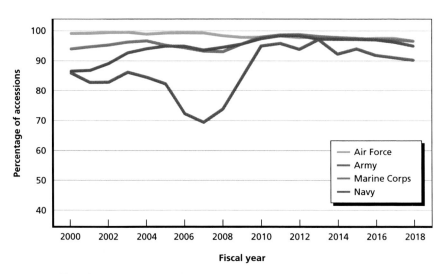

SOURCE: Office of People Analytics, undated.
NOTE: An HSDG is someone with a high school diploma and not exclusively a GED, associate's
degree, professional nursing diploma, bachelor's degree, master's degree, post–master's degree,
first professional degree, doctoral degree, post-doctorate work, or one semester of college completed.

increased in the other services.[2] Note that the fall in quality among
Army accessions was driven by both a reduction in HSDGs and a
smaller decrease in the percentage of overall recruits who scored in category IIIA or above (as shown in Figure 3.3).

Raw trends in quality of recruits over time do not account for
other factors that were also changing over this time period such as the
outside job options. We used the following explanatory variables in
the reduced form regressions described below to isolate the effect of
pay and better control for these other factors: military and civilian pay,
recruiting goal, deployment, unemployment, gender, and a post-2009
indicator (that is, an indicator being after 2009) to control for changes
in educational benefits policy. We describe the purpose of the latter
variable in greater detail later in this section.

---

[2]   For some ideas about why the Army failed to increase its quality at a time of increasing
RMC, see Hosek et al. (2018, pp. 71–73).

Figure 3.3
Percentage of Active-Component Non–Prior Service Accessions Who Were in Categories I–IIIA, by Service, FY 2000–2018

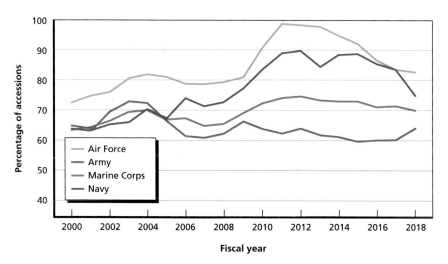

SOURCE: Office of People Analytics, undated.
NOTE: Category I–IIIA personnel are those who scored in the upper half of the AFQT score distribution.

To construct a variable for military and civilian pay, we used RMC for an E-4 with four YOS and the median civilian wages of 18- to 22-year-old male and female workers with high school education (and not more; including a General Education Development [GED] certificate). As a measure of pay we used the military/civilian wage ratio. This was constructed from data from the Greenbook and from the March CPS. We considered two pay measures—the RMC percentile and the military/civilian wage ratio—and chose the latter, which is consistent with the approach used in past studies (e.g., Simon and Warner, 2007; Asch et al., 2010). The former approach has the disadvantage that pay must increase by a larger absolute amount to move from, say, the 85th to the 90th percentile than from, say, the 70th to the 75th, while the pay ratio has the same interpretation throughout its range. That said, the regression results using the RMC percentile were nearly the same as the pay ratio results and had similar statistical significance. RMC is from the Greenbook, and the civilian wage is from the March CPS. We chose RMC of an E-4 at four YOS to approximate pay at the end of the first enlistment term.

In Figures 3.4 and 3.5 we show the RMC percentile, smoothed to adjust for variation resulting from sample size and the RMC/wage ratio.[3] Since females in the civilian sector make less on average than their male counterparts, the RMC percentile is higher for women than for men. Figure 3.5 shows the raw ratio and a linear curve fitted to it.

Figure 3.6 shows the recruiting goals for each service over the same time period. The Army had the largest recruiting goal, roughly double that of each of the other services in the middle of the decade. The recruiting goals decreased on net during this period, but the timing of the decrease differed by service. Recruiting goals have generally increased in recent years.

Figure 3.4
Smoothed Regular Military Compensation Percentiles: Male and Female High School Graduates, Ages 18–22, FY 1999–2017

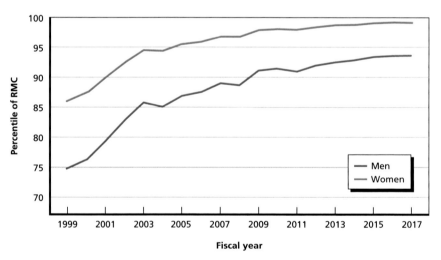

SOURCES: Active-duty pay files from DMDC; U.S. Census Bureau, 2018.
NOTE: We estimated the curves as described in Appendix A. The data used in the regressions were based on our tabulations of RMC as a percentile of the civilian wages of male and female high school graduates ages 18–22. RMC is for an E-4 with four YOS. The RMC percentile is relative to the wage distribution for 18- to 22-year-old workers with high school (and not additional) education who had more than 35 hours of work in the year and more than 35 usual weekly hours of work.

---

[3]   Appendix A describes the smoothing method and contains the raw and smoothed values.

Figure 3.5
Regular Military Compensation/Median Wage Ratio: Male and Female High School Graduates, Ages 18–22, FY 1999–2017

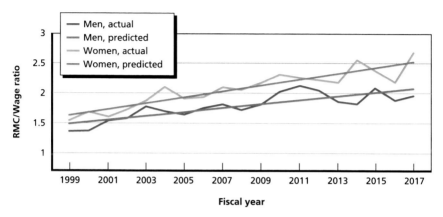

SOURCES: Active-duty pay files from DMDC; U.S. Census Bureau, 2018.
NOTE: We estimated the curves as described in Appendix A. The data used in the regressions were based on our tabulations of RMC as a percentile of the civilian wages of male and female high school graduates ages 18–22. RMC is for an E-4 with four YOS. The RMC percentile is relative to the wage distribution for 18- to 22-year-old workers with high school (and not additional) education who had more than 35 hours of work in the year and more than 35 usual weekly hours of work.

Figure 3.6
Recruiting Goals, by Service, FY 1999–2017

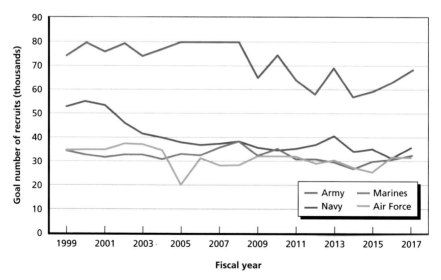

SOURCE: Accession Policy Directorate, 2017.

To get a measure of deployment, we use the number of person-nel receiving imminent-danger or hostile-fire pay. With 1999 normal-ized to one for each service, the Army and Marine Corps had 5 to 11 times more deployed personnel between 2003 and 2010 than in 1999 (Figure 3.7). Navy and Air Force deployed personnel were .9 to 2.5 times their 1999 level.

Extensive deployments between 2002 and 2009 might have made it more difficult for the Army to enlist high-quality recruits. The per-centage of Army accessions who were HSDGs dropped in the middle of the decade (Figure 2.2), which also meant a drop in high-quality accessions (Figure 2.1). Marine Corps deployment also rose, but the service's HSDG and high-quality recruiting rose as well.

Figure 3.8 shows changes in unemployment during this time period. When the economy worsens, workers have fewer outside options and may be more likely to be enticed to join the military, which pro-vides relatively high and stable pay. The percentage of high-quality recruits who were in categories I through IIIA rose rapidly from 2009

Figure 3.7
Enlisted Personnel Receiving Imminent-Danger or Hostile-Fire Pay,
Calendar Years 1999–2017

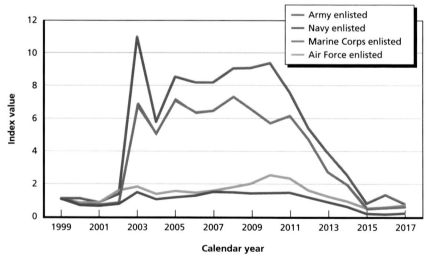

SOURCES: Author's calculations based on Defense Manpower Data Center Active-Duty Pay files.
NOTE: 1999 = 1.00.

Figure 3.8
Civilian Unemployment Rate, Calendar Years 2000–2018

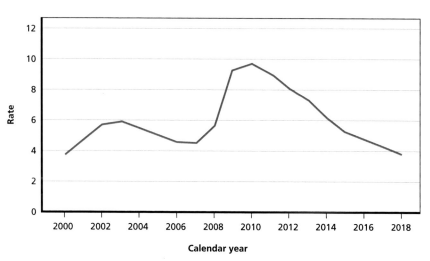

SOURCE: Bureau of Labor Statistics, undated.

to 2012, a period when the unemployment rate was high relative to 2000–2008 and 2013–2018. Researchers in past studies have found that the number of high-quality enlistments in each service, including the Army, is positively associated with the civilian unemployment rate (for reviews of past studies, see, e.g., Asch, Hosek, and Warner [2007]; Asch et al. [2010]).

We also include a post-2009 indicator variable to capture the changing nature of educational benefits for service members in different branches. The Post-9/11 Veterans Educational Assistance Act of 2008 (also known as the Post-9/11 GI Bill) for education benefits took effect in August 2009 (Pub. L. 110-252, Title V, 2008). This bill covered tuition at a level equal to the tuition of a service member's home-state four-year public university, plus offering BAH while attending school. Prior to that point, while the services had the same base education benefit, the Army offered a supplement, the Army College Fund, for high-quality recruits in critical occupations. The Post-9/11 GI Bill made benefits available to recruits of all services on equal terms, regardless of quality, and, given the generosity of the benefits, the Army lost the recruiting advantage it had.

Modeling the Relationship Between Recruiting Rate and Regular Military Compensation/Wage Ratio

Recruiting is determined both by the willingness of an individual to enlist and by the service's willingness to accept the recruit. The model we estimated is a reduced-form model that reflects supply-and-demand influences but does not identify the effects of the supply-and-demand variables separately.

We view willingness to enlist as a variant of the random utility model (McFadden, 1983). An individual's willingness to join a service depends on military pay relative to civilian pay, job opportunities as measured by the unemployment rate, job and school opportunities related to AFQT, the chance of being deployed in hostile operations, and possible differences in preferences and opportunities related to gender. It also depends on factors not observed in our data, including the military occupational specialties that are offered, bonuses, educational benefits, ship date, information from advertising or service websites, the influence of family and friends, and aspects of military service, such as its roles, missions, tradition, and values (see, e.g., Eighmey, 2006).

On the demand side, the recruiting command wants to meet a quantity goal and meet or exceed a quality goal. The service goals must comply with DoD guidance that calls for at least 60 percent of accessions to be in categories I through IIIA and at least 90 percent to be tier 1 (Kapp, 2013; Sellman, 2004). A tier 1 recruit is one with a high school diploma or at least 15 college credits; tier 1 recruits are predominantly HSDGs. Category I–IIIA HSDG recruits count toward the quantity and quality goals, while category IIIB HSDG and category IV HSDG recruits count only toward the quantity goal. This suggests that the probability of an HSDG's recruitment conditional on the person being willing to enlist relates to a preference ordering:

$$Pr(accept|willingCatsIthroughIIIA) = 1$$
$$Pr(accept|willingCatIIIB) \leq 1$$
$$Pr(accept|willingCatIV) \ll 1$$

The "equal" part of "less than or equal" for IIIB allows for the possibility that meeting the quantity goal might require accepting all willing IIIBs. It also reflects the possibility of identifying high-potential IIIB prospects through nontraditional methods. Thus, the AFQT can affect a service's willingness to accept someone who is willing to join, especially if the person is at IIIB or less. Also, a person's willingness can depend on AFQT results, because, as mentioned, these can be related to job prospects, college expectations, and college opportunities. For these demand-and-supply reasons, our analysis allowed for possible differences by AFQT category in the recruiting rate's responsiveness to the RMC/wage ratio.

Our model is a reduced-form model because it does not identify structural equations for the demand-and-supply sides. For instance, unemployment can increase willingness to enlist, and in response the services can decrease their recruiting resources, such as recruiters, advertising, and bonuses, and tighten eligibility. Our reduced-form unemployment coefficient is the net effect of these responses. The direct effect of unemployment on supply is positive, but if unemployment triggers a large enough decrease in recruiting resources at the same time, the coefficient on unemployment in the reduced-form model could be negative. Similarly, deployment might have a positive or negative effect on supply, and if the supply effect is negative, the service might take the option of increasing its enlistment bonuses (as mentioned above) to nullify the negative effect.

It would be ideal to identify structural (causal) effects of recruiting resources, which are endogenous to the recruiting process, but doing so requires exogenous variation in explanatory variables, such as through enlistment bonus, recruiter, or advertising experiments, or through the use of instrumental variables. Our data are not from experiments, and we did not have instruments or enough data to estimate instrumental variable models. Instead, the variables we included in the reduced-form model were outside the control of the recruiting command and, as suggested, are external to its resourcing decisions. These variables are military and civilian pay, recruiting goal, deployment, unemployment, and eligibility. Although the reduced-form model does not identify causal effects, such as the causal effect of pay, it avoids issues of bias that would have arisen if we had included observed bonuses, recruiters, and advertising.

Still, poor recruiting and retention conditions in one year might result in a higher-than-expected military-pay raise and a higher recruiting goal in the next. These conditions—autocorrelated errors and policy actions affecting pay and retention—could bias downward the RMC/wage and recruiting goal coefficients and produce low standard errors (and thus high t-statistics). A downward bias would imply that the coefficients are conservative estimates of the true effect.

We ran separate models for each service. The dependent variable is the logit of the recruiting rate. There is a recruiting-rate observation for each AFQT category for men and another for women. The RMC/wage coefficient is allowed to differ by category. Also, the intercepts are allowed to differ by AFQT category interacted with gender. The coefficients for recruiting goal, deployment, unemployment, and the post-2009 indicator are the same across AFQT categories. In the logit specification, the *percentage* change in the recruiting rate with respect to a continuous variable equals the coefficient times one minus the recruiting rate (i.e., $\beta[1 - p]$).[4] The recruiting rates are low percentages, so $1 - p \approx 1$, and the percentage change is roughly equal to the coefficient itself, β.[5]

---

[4]   In the logit regression specification, the marginal change in p with respect to a continuous explanatory variable x is

$$\left(\frac{\partial p}{\partial x}\right) = \beta(1 - p)p.$$

Therefore, the percentage change in p with respect to a one-unit change in x is

$$\left(\frac{\partial p}{\partial x}\right)\left(\frac{1}{p}\right) = \beta(1 - p).$$

We also ran regressions in which the dependent variable was the log of the recruiting rate. In that specification, the coefficients represent the percentage change in the dependent variable for a one-unit change in the explanatory variable. The results were virtually the same as $\beta(1 - p)$ from the logit.

[5]   For indicator variables, the impact on the recruiting rate required evaluating

$$\frac{e^{\beta x}}{1 + e^{\beta x}}$$

for the variable at 1 versus 0, with other explanatory variables held at some level (e.g., their means).

A final point is that one might expect the results to differ for the Army and the other services because the Army's recruiting goal is the largest, being roughly twice that of each of the other services. However, the extent to which the magnitude of the recruiting goal makes a difference depends on resourcing decisions that the services made and the quality they required. The marginal cost of a recruit of a given quality might be higher for the Army if it must go deeper into the population of prospective recruits; yet, by programming enough resources to recruiting, the Army might attain the same quality as the other services. But resources have other uses, and the same quality might not be needed. The issue, then, turns on the expected benefit from higher-quality recruits relative to their cost and, further, whether the positions in the Army need to be manned by the same quality of recruit, on average, as needed in the other services. In short, service differences in recruiting cost and required quality could give rise to differences in recruit quality and—relevant to this research—differential responses in recruit quality by service when the rise in RMC is the same across services.

Modeling the Relationship Between Share of Non–High School Diploma Graduate Accessions and Regular Military Compensation/Wage Ratio

Here, we focus on the share of accessions who are non-HSDGs, again by AFQT category. The intuition is that the service will prefer an HSDG recruit to a non-HSDG recruit, other things being equal. Yet, if recruiting conditions are more difficult than expected, recruiting outcomes are below goal, goals have unexpectedly increased, or recruiting stations become short-staffed, recruiting a non-HSDG might be more attractive. This would be the case if, given the adverse recruiting conditions, the marginal benefit from meeting the quantity goal relative to the marginal cost of doing so were greater for a non-HSDG than an HSDG recruit. This intuition extends beyond unexpected, difficult recruiting conditions, though. Judging by recruiting outcomes (Figures 3.1 through 3.3), non-HSDG accessions are a regular part of the recruiting mix that the services program resources to obtain, although the actual outcomes will depend on conditions realized during the recruiting year. In 1999, more than 10 percent of Army and Navy accessions were non-HSDG, and, although the percentage in recent years has been well below 10 percent for all services, in 2006 and 2007, more than 35 percent of male Army accessions were non-HSDGs. Overall, the cost of an all high-quality or all-HSDG accession cohort might be too high relative to the expected value to the service.

To capture the idea that non-HSDG accessions serve as an outlet, we estimated models of the share of accessions who were non-HSDGs. The models include the same explanatory variables as the recruiting rate models, and the dependent variable is the logit of the share of accessions who were non-HSDGs.

Regression Results

The regression estimates are reported in Tables B.6 through B.9 in Appendix B. We summarize our findings in Figures 3.9 and 3.10. These figures depict the estimated coefficient for the ratio of RMC to civilian wage, by AFQT category for each service for the two models, respectively. Thus, the figures show the estimated relationship between the RMC/wage ratio and the recruiting outcome variable by AFQT category for each service.

Recruiting Rate
As shown in Figure 3.9, an increase in the RMC/wage ratio was associated with

- no change in the recruiting rate for category I for the Army but decreases in the rates for II, IIIA, and IIIB
- an increase for the Navy in recruiting rates for categories I and II
- increases in all categories—I, II, IIIA, and IIIB—for the Marine Corps

Figure 3.9
Regular Military Compensation/Wage Coefficients from Logit Regressions of Recruiting Rate for Armed Forces Qualification Test Categories I–IIIB, by Service

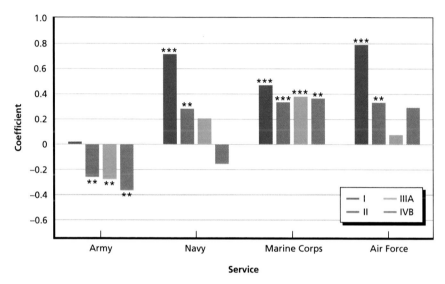

NOTE: The bars in the figure show the overall RMC percentile coefficient for each AFQT category. For example, the overall RMC percentile coefficient for category I is RMC's effect on the logit recruiting rate for AFQT category I. It equals the sum of the RMC coefficient for IIIA, the reference group, and RMC × category I coefficient in Tables B.6–B.7.
*** = statistically significant at the 1-percent level; ** = statistically significant at the 5-percent level; * = statistically significant at the 10-percent level.

- an increase for the Air Force in recruiting rates for categories I and II.

Thus, the Navy and Air Force increased their quality mix through higher category I and II recruiting rates over the time period. The Marine Corps also increased recruiting rates for categories I, II, and IIIA (as well as IIIB). Like Hosek et al. (2018, pp. 52–63), we find that an increase in the RMC/wage ratio for the Army was associated with no change in the category I recruiting rate and, contrary to what one might expect, lower recruiting rates in II and IIIA, as well as IIIB.

Figure 3.10
Regular Military Compensation/Wage Coefficients from Logit Regressions of the Non–High School Diploma Graduate Share of Accessions for Armed Forces Qualification Test Categories I–IIIB, by Service

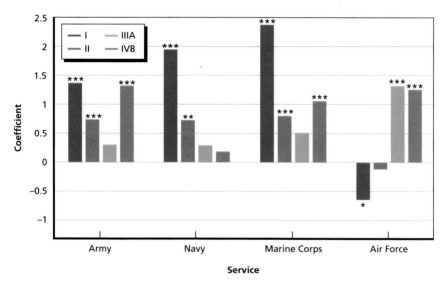

NOTE: The bars in the figure show the overall RMC percentile coefficient for each AFQT category. For example, the overall RMC percentile coefficient for category I is RMC's effect on the share of non-HSDG accessions for AFQT category I. It equals the sum of the RMC coefficient for IIIA, the reference group, and RMC × category I coefficient in Tables B.7–B.9.
*** = statistically significant at the 1-percent level; ** = statistically significant at the 5-percent level; * = statistically significant at the 10-percent level.

In short, increasing RMC relative to civilian wages was associated with an increase in the recruiting rate, especially in categories I and II for all services except the Army. Hosek et al. (2018, pp. 71–73) discuss in depth why the results for the Army might be different than for the other services. It could be that Army recruiting became more difficult for reasons not captured in our models. Another possibility is that the Army set quality goals and programmed recruiting resources to sustain, but not increase, accession quality or that higher RMC affected recruiter effort. Yet another explanation is that Army recruiters were allowed to reduce effort as the recruiting market expanded as RMC increased, so recruiting rates in II and IIIA were lower. Hosek et al. (2018) explicitly model a trade-off between RMC and a service's recruiting resources, including recruiter effort, and show that such a model could explain these results.

Non–High School Diploma Graduate Share

As shown in Figure 3.10, an increase in the RMC/wage ratio was associated with

- an increase for the Army in the share of non-HSDG in categories I, II, and IIIB
- an increase in the share of non-HSDG in categories I and II for the Navy
- an increase for the Marine Corps in the share of non-HSDG in categories I, II, and IIIB
- an increase in the RMC/wage ratio associated with a decrease in the share of non-HSDGs in category I and an increase in the share in categories IIIA and IIIB for the Air Force.

Controlling for the factors discussed above, we found that an increase in the RMC/wage ratio was associated with a higher share of non-HSDGs across all services and with a higher share of non-HSDGs in categories I, II, and IIIB for the Army and Marine Corps and in categories I and II in the Navy.[6] Thus, in the Army, Navy, and Marine Corps, the RMC/wage ratio was associated with an increase in the quality of non-HSDGs. Still, the non-HSDG share in IIIB also increased in the Army and Marine Corps. The findings for the Air Force were mixed, with a decrease in this share for category I but an increase in the share in IIIA and IIIB.

Conclusion

This chapter has analyzed the relationship between military pay and recruit quality. While the RMC/wage ratio for an E-4 with four YOS

---

[6] With no controls, the relationship between share of non-HSDG and RMC/wage ratio is negative for all branches except the Air Force.

rose significantly during this period, recruit quality varied by service. We used a reduced form regression analysis to control for several factors and shed further light on the association between RMC and recruit quality.

As the regressions indicate, recruit quality increased for all services except the Army as the RMC/wage ratio increased. The Navy and Air Force increased their category I and II recruiting rates; and the Marine Corps increased its category I, II, and IIIA recruits (as well as IIIB recruits). In contrast, the association between the RMC/wage ratio and the Army recruiting rate for II, IIIA, and IIIB was negative.

We also found that a higher RMC/wage ratio was associated with a higher share of non-HSDG accessions in categories I, II, and IIIB for the Army and Marine Corps, categories I and II for the Navy, and categories IIIA and IIIB for the Air Force. A higher RMC/wage ratio was correlated with a decrease in share of non-HSDG accessions in category I for the Air Force.

Our regression models are reduced form and did not identify the causal effect that military or civilian pay has on recruiting outcomes. Ideally, each of our explanatory variables would be varied experimentally in order to avoid endogeneity. However, the military does not have direct control over many of our explanatory variables, and in the past, it has been reluctant to vary policy levers that it does control (such as bonuses, advertising, and recruiting locations), because the importance of meeting recruiting missions has often taken precedence over experimentation.

CHAPTER FOUR

Geographic Differences in Regular Military Compensation Percentiles

Pay comparisons in past QRMC reports as well as in the comparisons shown in previous chapters in this report are based on national data. That is, RMC is compared with the pay of civilians throughout the United States, without regard to where individuals live. Such national comparisons are useful because they provide a summary measure of the overall status of military pay relative to civilian pay and because basic pay, the foundation of RMC, is set at the national, not the regional, level. Further, personnel rotate frequently and are not attached to a specific location over their entire career, so from the standpoint of the retention decisions of military personnel, comparison of military pay with pay across the national external market seems relevant. The relevance of the national market is less clear for the initial recruiting decision since potential recruits are likely to put more weight on how military pay compares with pay in their hometown region than with pay at the national level. In general, more research is needed to better understand the extent to which pay comparisons at the local rather than the national level influence both recruiting and retention decisions and the extent to which pay supplements such as BAH help to make pay competitive in different areas.

Insofar as local pay comparisons are important for these decisions, it is useful to consider geographic differences in RMC percentiles, especially because the structure of wages in urban versus non-urban regions of the country has shifted in recent years and thereby affected how RMC compares with civilian pay in urban versus non-urban areas. In this chapter, we highlight these comparisons.

Researchers are becoming increasingly aware of the important ways in which location influences life outcomes. Several studies have examined the influence of neighborhoods on wages and social mobility (Chetty and Hendren, 2018a, 2018b; Chetty, Hendren, and Katz, 2016), the importance of state laws for promoting economic growth or access to healthcare (Clarke, 2004; Currie and MacLeod, 2008; Holmes, 1998; Kleiner, 2016), and the differences in economic opportunities between cities and more rural areas (Glaeser, 2010, 2011). However, recent research has also documented changes in the complex interactions between educational attainment, economic opportunities, and geography (Autor, 2019; Austin, Glaeser, and Summers, 2018). In previous decades, cities offered wage premiums to both highly skilled and less-skilled individuals. That is, pay was higher in cities than outside cities for both highly skilled and less-skilled workers. Urban areas had dynamic labor markets supporting large manufacturing facilities as well as specialized high-skilled innovation. For example, Detroit became known for large car manufacturing plants. Relatively high-paying jobs attracted workers who then specialized in the skills necessary for their work. In turn, firms in adjacent industries had an incentive to locate to Detroit to take advantage of this large pool of skilled workers. After gaining experience in the industry, workers were also more likely to start new firms, which further attracted additional workers to the area. These benefits of geographic concentration are called *agglomeration effects*.

In previous decades, the higher productivity and higher wages from agglomeration effects encouraged more people to move to cities. However, in the last several years many of the relatively high-paying jobs that were previously done by individuals with less formal education have been automated or outsourced, and many of these workers have moved into lower-paying service-sector jobs (Acemoglu and Autor, 2011; Acemoglu and Restrepo, 2017, 2018; Alabdulkareem et al., 2018; Autor, 2015, 2019; Autor and Dorn, 2013; Autor, Katz, and Kearney, 2006; Autor, Levy, and Murnane, 2003). The result is that wages are now no higher in more urban areas for individuals with less formal education than they are in less urban areas. For highly educated workers, there is still a large wage premium for moving to cities. At the same time, other policies (such as limitations on new construction, increases in local taxes, and licensing requirements) have increased the costs of living in many cities even as the premium for workers with less formal education has disappeared (Ganong and Shoag, 2017).

Figure 4.1 illustrates the result of this trend. It shows median weekly wages from the CPS for 2014–2017 for those who have different levels of education and who live in the ten least urban states and in

Figure 4.1
Weekly Wages by Education Level for Those in the Ten Least Urban States and Those in the Ten Most Urban States, 2014–2017

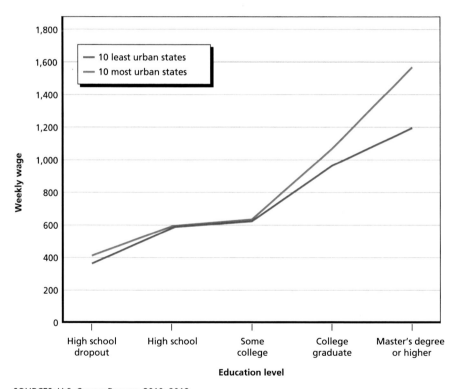

SOURCES: U.S. Census Bureau, 2010, 2018.
NOTE: Median weekly wages are in 2017 dollars.

the ten most urban states.[1] As can be seen, workers with lower levels of education make around the same amount whether they live in more or less urban areas. Those with higher levels of education make significantly more in more urban states.

To the extent that pay at the local level is relevant to recruiting and retention decisions, the implication of this trend in the civilian labor market is that military service was less competitive in urban than in less urban areas in the past because pay was higher in urban areas for both higher- and lower-skilled workers. Because of the shifts in the urban labor markets for lower-skilled workers over the last few decades, it seems likely that military service may now be just as competitive for lower-skilled workers in urban as in less urban areas, all else being equal. The lower-skilled workers include those with high school and some college, as shown in Figure 4.1, and are a prime market for the enlisted force, especially the junior enlisted force.

Geographic Differences in Regular Military Compensation Percentiles for Enlisted and Officers

To some extent RMC includes a geographic component. As an element of RMC, BAH provides an adjustment to pay based on the cost of housing in different areas. Also, the services offer enlistment and reenlistment bonuses to eligible individuals, and such bonuses help make military service more competitive to the extent that RMC does not compare as favorably in a specific area. Still, it is useful to consider

---

[1]   States are categorized based on the percentage of the total population in urban areas from the 2010 Decennial Census. The District of Columbia, although not a state, is included in the measure since more people live there than many states. Note that the definition is the percentage of the population who lives in urban areas and not the percent of the total land mass filled by people. Thus, states that we do not often think of as urban because they have lots of empty land are surprisingly dense because most of the population lives in a small geographic area. This is the case with states such as Utah, Nevada, and Arizona. The ten densest states are District of Columbia, California, New Jersey, Nevada, Massachusetts, Hawaii, Florida, Rhode Island, Utah, and Arizona. The ten least urban states are North Dakota, Alabama, Kentucky, South Dakota, Arkansas, Montana, Mississippi, West Virginia, Vermont, and Maine.

how the changing dynamics of civilian wages by geography affect how RMC compares with civilian wages in urban and less urban areas.

Figures 4.2 and 4.3 summarize the regional comparisons shown in Appendix C by showing RMC for enlisted and officers, respectively, as the gray line graphed relative to the right axis. On the left axis, the figures show the percentile values along the RMC line, by year of service/age. The figures show how RMC wages compare with civilian wages over a career in the most urban and the least urban states.

For enlisted personnel in Figure 4.2, RMC is a higher percentile of civilian wages compared with civilians who live in the most urban

Figure 4.2
Enlisted Regular Military Compensation and Regular Military Compensation Percentiles for Full-Time, Full-Year Workers with High School Diploma, Some College, and Associate's Degree in the Most and Least Urban States, 2017

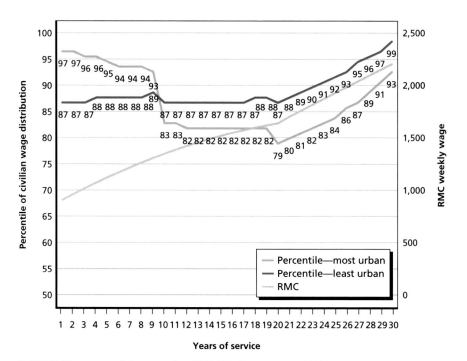

SOURCES: Directorate of Compensation, 2017; U.S. Census Bureau, 2018.
NOTE: RMC percentile varies by YOS (1–9 = high school, 10–19 = some college, and 20–30 = associate's degree). We weighted civilian-wage data by enlisted military gender mix. The gray line is enlisted RMC. Data are smoothed.

Figure 4.3
Officer Regular Military Compensation and Regular Military Compensation Percentiles for Full-Time, Full-Year Workers with Bachelor's Degree or with Master's Degree or Higher in the Most and Least Urban States, 2017

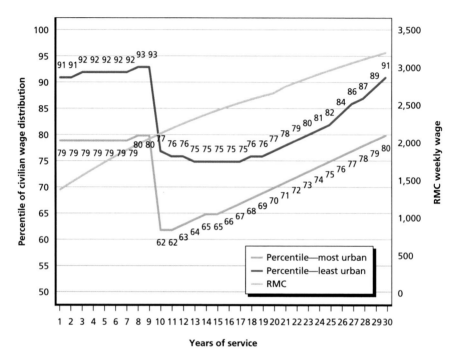

SOURCES: Directorate of Compensation, 2017; U.S. Census Bureau, 2018.
NOTE: RMC percentile varies by YOS (1–9 = bachelor's degree, 10–30 = master's degree or higher). We weighted civilian-wage data by officer military gender mix. The gray line is enlisted RMC. Data are smoothed.

versus the least urban states for one through nine YOS. That is, for enlisted personnel early in their career, RMC compares more favorably with civilian pay in urban areas than in nonurban areas. This pattern flips as we compare RMC with the pay for civilians with more years of formal education as years of service increase. However, the difference between the percentiles in the least and most urban states is about 5 percentiles (e.g., 87th percentile for less urban workers and 82nd percentile for more urban workers for 12–19 YOS) and grows to about 8 percentage points at 20 YOS. That is, for enlisted personnel in their later career, RMC compares somewhat less favorably with civilian pay in urban areas than in non-urban areas. But, notably, the 5 to 8 percentile differences after ten YOS are smaller than the differences for officers, which are shown in Figure 4.3

In Figure 4.3 we compare RMC for officers from the Greenbooks with civilians with bachelor's degrees (1–9 YOS) and master's degrees or more (10–30 YOS). While exhibiting a similar overall pattern to that found in Figure 2.2, officer RMC is at a persistently and substantially higher percentage of civilian pay in the least urban states than in the most urban states; in most cases the difference is more than ten percentiles. That is, because of the higher wages of urban civilian workers, RMC percentiles are lower in urban areas than in less urban areas by more than ten percentiles.

Traditionally, wages have been lower in less urban areas so we would expect RMC to be a higher percentile of civilian wages in these areas. This is still true for workers with more years of formal education; RMC compares more favorably with what workers with more years of formal education in less urban states could get in their local labor market. However, for less-educated workers, the difference between wages in more and less urban areas is much smaller; and for those with only a high school degree, wages may now be higher in less urban areas, thereby making RMC a higher percentile of civilian wages in more urban areas.

Conclusion

There appear to be differences in RMC percentiles across geographies for individuals of different education levels. Due to trends for less-educated workers across the economy, enlisted military members with a high school degree are more likely to find RMC equally competitive no matter where they live (and perhaps even more competitive in urban areas). However, officers and those with more education in general are likely to find military pay more competitive if they live in less urban areas than if they live in more urban areas.

While these patterns are briefly noted here, further research is necessary to explore their implications for military recruiting and retention.

CHAPTER FIVE

Closing Thoughts

Our findings are relevant to the following questions. How does military pay for active-component personnel compare with civilian pay? Has the position of military pay improved or worsened since 2009, when the 11th QRMC last benchmarked military pay? Given that military pay has increased since 1999, when the 9th QRMC first benchmarked military pay, was that increase associated with an increase in recruit quality? We summarize our findings and offer some final thoughts.

Findings in Brief

At What Percentile Did Regular Military Compensation Stand in 2017?

We find that RMC in 2017 was at the 85th percentile of comparably educated civilian wages for active-component enlisted personnel and at the 77th percentile for active-component officers. The 9th QRMC noted that many enlisted members have some college and recommended that military pay be at around the 70th percentile for that level of education. In 2017, we find that RMC was at the 84th percentile for enlisted with some college and the 93rd percentile for those with high school. For officers, RMC was at the 86th percentile for officers with a bachelor's degree and the 70th percentile for officers with a master's degree or higher. We also compared RMC with civilian wages over time from 2000 to 2017 for selected age and education groups. These comparisons showed a steady increase in RMC relative to civilian pay from 2000 to 2010 and a leveling off afterward.

At What Percentile Did Regular Military Compensation Stand in 2009?

RMC in 2009 was at the 84th percentile of civilians comparable with enlisted members and at the 77th percentile of civilians comparable with officers—the same as 2017. RMC was at the 85th percentile for enlisted with some college and the 91st percentile for those with high school. RMC was at the 87th percentile for officers with a bachelor's degree and the 69th percentile for those with a master's degree or higher. Our RMC percentiles for 2009 are somewhat below those of the 11th QRMC estimates—90th percentile for enlisted personnel and 83rd for officers—and we attribute the difference to methodological considerations described in Hosek et al. (2018, pp. 10–16, 28–30).

Our finding that the RMC percentile was nearly the same in 2017 as 2009 might be surprising because there were years when basic-pay raises were below the Employment Cost Index (ECI)—namely, 2014–2016—though in our view the ECI is not a reliable guide for military and civilian pay comparisons for various reasons, including that it is not adjusted for the military education distribution.

How Did Recruit Quality Change as Regular Military Compensation Rose?

We used regression models to isolate the relationship between the RMC/wage ratio and recruiting outcomes. We found that as the RMC/wage ratio increased, recruit quality increased in the Navy, Marine Corps, and Air Force but not in the Army.

The regressions controlled for recruiting goal, deployment, unemployment, and gender and were estimated separately by service. As the RMC/wage ratio rose, the Navy, Marine Corps, and Air Force increased their AFQT category I and II recruiting rates, and the Marine Corps also increased its category IIIA rate. But the Army decreased its category II and IIIA recruiting rates. The Army, Navy, and Marine Corps increased the percentage of non-HSDG accessions in category I and II while the Air Force reduced the percentages of non-HSDG accessions in category I. These results suggest that the Army, Navy, and Marine Corps increased the quality of their non-HSDG recruits.

The reason for the Army's different result is an open question. It is possible that Army recruiting became more difficult during the 2000s because of extensive deployments in support of operations in Iraq and Afghanistan, and the Army did not program enough recruiting resources to match the increased difficulty. It may also be that the Army set its recruiting quality goals to hold recruit quality constant as RMC increased, thereby holding its recruit quality near the DoD quality benchmarks of at least 90 percent HSDG recruits and at least 60 percent from categories I through IIIA, rather than allocating more resources such as bonuses and recruiters to recruiting.

How Does the Regular Military Compensation Percentile Vary Across Geographies Within the United States?

There appear to be large differences in how RMC compares with civilian pay across geographies for individuals of different education levels. Due to trends for less-educated workers across the economy, enlisted military members with a high school degree are likely to find military pay relative to civilian pay as attractive in urban as in nonurban areas—a change from earlier years when military pay was relatively less attractive. However, officers and those with more education in general are likely to find military pay higher relative to civilian pay if they live in less urban areas than if they live in more urban areas.

Wrap-Up

In short, our results indicate that RMC continues to exceed the 70th percentile of the distribution of pay of civilians with similar characteristics, found by the 11th QRMC and more recently in Hosek et al. (2018). This result is reached even when accounting for an increase in the educational attainment of both the enlisted force and officers. Our analysis indicates that since 2000, increases in RMC relative to civilian pay have been associated with increases in recruit quality, with the exception of the Army.

These results raise some additional questions. First, the defense capability gained from recruiting more high-quality recruits must be weighed against the added cost of higher RMC, which increases the entire personnel budget. Our analysis has shown where and to what extent recruit quality increased as military pay increased, but it cannot place a value on the increased quality. Valuing quality is the services' domain.

Second, the increase in recruit quality for three of the four services between 1999 and 2017 raises the question of whether achieving that quality could have been accomplished in a more cost-effective manner than increasing RMC relative to civilian pay. Though estimates differ on the marginal cost of pay and recruiting resources, virtually every study finds that military pay is the costliest approach for enlisting high-quality recruits (Asch et al., 2010; Orvis et al., 2016; Simon and Warner, 2007). RMC is a blunt instrument that is not targeted to occupational specialties in which recruiting or retention shortfalls occur. An increase in RMC affects the cost of all personnel budget in every service, while an increase in a service's recruiting resources such as recruiters, enlistment bonuses, advertising, and recruiting stations and equipment is specific to its recruiting budget, and resources such as bonuses can be targeted.

These questions should be addressed when considering the setting of RMC in the future. The analysis in this report indicates that in the recent past, and since the 11th QRMC, RMC continues to support readiness and lies above the benchmark of the 70th percentile set by the 9th QRMC.

Regular Military Compensation Percentile and Regular Military Compensation/Wage Ratio

This appendix describes how we smoothed the RMC percentile for an E-4 with four YOS relative to the civilian-wage distribution for high school graduates ages 18 to 22 who worked more than 35 weeks in the year and had more than 35 usual hours of work. The smoothed values are shown in Figure 3.5 in Chapter Three. This appendix also describes the ratio of RMC to the median civilian wage. RMC is based on Greenbook data, and civilian wages come from March CPSs.

Smoothing the Regular Military Compensation Percentile

Raw RMC percentiles vary considerably from year to year. Some variation comes from annual increases in RMC resulting from increases in basic pay and BAH, but much of the variation comes from the smallness of CPS samples. For high school graduates ages 18 to 22, sample sizes for each year of data range from 150 to 250 observations for men and the same for women. These sample sizes do not provide dense enough coverage for a precise estimate of the wage distribution or the RMC percentile. We used a smoothing method to adjust for the variation.

There are different approaches to smoothing. One approach is a kernel density estimator to smooth the wage distribution each year. But that approach does not use data from adjacent years and cannot be relied on to provide year-to-year continuity. Instead, we estimated a log wage model to identify the mean and variance of the wage distribution, allowing for a common trend in real wages. The estimation used the tabulated wages at the 30th, 40th, 50th, 60th, 70th, 80th, and 90th percentiles. Then, using the estimated wage distribution parameters, we inferred the RMC percentile. The approach smooths the RMC percentile and provides year-to-year continuity in the wage distribution as real wages change over time.

Using Wage Percentiles to Estimate the Log Wage Distribution

Let p be the percentile (e.g., $p = 0.6$ at the 60th percentile), and let F be the standard normal distribution with mean 0 and standard deviation σ. Assume that the wage is log-normally distributed, so

$$p = F \left[\frac{\ln w_p - \mu}{\sigma} \right]. \tag{A.1}$$

Taking the inverse normal, the log wage at percentile p is
$\ln w_p = \mu + \sigma F^{-1}[p]$.

We tabulated CPS wage data to find wages at the 30th through 90th percentiles for each year, 1999 through 2017. These wages are the observations on $\ln w_p$. Thus, for a given group (e.g., 18- to 22-year-old high school graduates), the log wage at percentile p in year t is

$$\ln w_{pt} = \mu_t + \sigma F^{-1}[p]. \tag{A.2}$$

At the 50th percentile, $F^{-1}[0.5] = 0$, and it follows that the mean of the log wage distribution, μ_t, equals the log of the wage at the 50th percentile. We computed the wage at the 50th percentile and used it to obtain an estimate of the mean of the log wage distribution: $\mu_t = \ln w_{0.5t}$ at each year and, in particular, for the base year of our data, 1999. We refer to 1999 as period 0.

For small changes, we approximated the year-to-year wage change as a percentage change from the wage in period 0:

$$w_{pt} = e^{\mu_0 + \sigma F^{-1}[p]} e^{\delta t}. \tag{A.3}$$

Taking logs, we have

$$ln \ w_{pt} = \mu_0 + \sigma F^{-1}[p] + \delta t. \tag{A.4}$$

We can replace μ_0 with $ln \ w_{0.5t} + \varepsilon_{pt}$ given that $w_{0.5,0}$ is computed from the data and its log is an estimate of the mean. Subtract $ln \ w_{0.5t}$ from both sides to obtain

$$ln \ w_{pt} - ln \ w_{0.5,0} = \sigma F^{-1}[p] + \delta t + \varepsilon_{pt}. \tag{A.5}$$

In the usual regression format, this can be thought of as

$$ln \ w_{pt} - ln \ w_{0.5,0} = \beta_1 F^{-1}[p] + \beta_2 t + \varepsilon_{pt}, \tag{A.6}$$

where β_1 is an estimate of σ, the standard deviation of the log wage, β_2 is an estimate of δ, the annual percentage change in the wage, and there is no intercept. Values for $F^{-1}[p]$ at each percentile came from the inverse normal function evaluated at the given percentile. The variable t is the year. This approach assumes that the standard deviation does not change during the observation period and that the mean evolves according to the time trend. For each group, there are wages for seven percentiles in each of 19 years, 1999 through 2017, for a total of 133 observations. There are two parameters to estimate.

Using the estimated parameters, the predicted wage in period t at percentile p is

$$w_{pt} = e^{ln \ w_{0.5,0} + \hat{\sigma} F^{-1}[p] + \hat{\delta} t}. \tag{A.7}$$

Also, for a given value of the wage, the corresponding percentile is derived as follows:

$$ln \ w_t - ln \ w_{0.5,0} = \hat{\sigma} F^{-1}[p] + \hat{\delta} t. \tag{A.8}$$

$$F^{-1}[p] = \left[\frac{ln \ w_t - ln \ w_{0.5,0} - \hat{\delta} t}{\hat{\sigma}} \right]. \tag{A.9}$$

$$p = F \left[\frac{ln \ w_t - ln \ w_{0.5,0} - \hat{\delta} t}{\hat{\sigma}} \right]. \tag{A.10}$$

Letting RMC_t stand in for wage, its percentile is

$$p = F\left[\frac{ln\ RMC_t - ln\ w_{0.5,0} - \hat{\delta}t}{\hat{\sigma}}\right]. \tag{A.11}$$

Parameter Estimates and Goodness of Fit

Table A.1 reports the regression results. The standard deviation of the log wage distribution is 0.457 for men and 0.398 for women, both of which are highly significant. Also, for interpretability, the estimates for δ are reported as the annual percentage change (i.e., as 100 times the estimated coefficient). For example, the δ estimate of −0.500 for male high school graduates reported in the table means that wages trended down by 0.500 percent per year from 1999 to 2017. Similarly, the reported standard error of δ is 100 times the estimated standard error. The wage data are in 2017 dollars. The trend estimate is statistically significant for men and women. The models fit the data well, with an R^2 of 0.93 for men and 0.88 for women.

Table A.1
Regression Results for High School Graduates, Ages 18–22

Group	Coefficient	Robust Standard Error	t	P>|t|
Male				
Standard deviation of log wage distribution	0.457	0.015	30.78	0.000
Time trend	−0.500	0.059	−8.42	0.000
R^2	0.929			
Female				
Standard deviation of log wage distribution	0.398	0.016	25.75	0.000
Time trend	−1.046	0.085	−12.28	0.000
R^2	0.883			

Raw and Smoothed Regular Military Compensation Percentiles

Table A.2 contains the raw and smoothed (predicted) RMC percentiles. In making predictions, we used the coefficients in Table A.1 and the median weekly wage for 1999, which was $565.89 for men and $500.25 for women.

Table A.2
Regular Military Compensation Percentiles for High School Graduate
Civilian Wages, Ages 18–22, as Percentages

Year	Males		Females	
	%Raw	%Smoothed	%Raw	%Smoothed
1999	79	75	94	86
2000	76	77	88	88
2001	81	79	86	90
2002	80	83	90	92
2003	89	86	96	94
2004	88	85	91	94
2005	88	87	95	95
2006	87	88	91	96
2007	92	89	99	97
2008	94	89	91	97
2009	90	91	94	98
2010	96	91	100	98
2011	94	91	100	98
2012	89	92	100	98
2013	82	92	88	98
2014	86	93	98	99
2015	90	93	99	99
2016	97	93	97	99
2017	87	93	99	99

SOURCES: Directorate of Compensation, 1999–2017; U.S. Census Bureau, 2018.
NOTE: RMC is for an E-4 with four YOS. Civilian wages are for 18- to 22-year-old workers with high school (and not additional) education who worked more than 35 weeks in the year and had more than 35 usual weekly hours of work.

The Regular Military Compensation/Wage Ratio

We again used RMC for an E-4 with four YOS and wages for civilians ages 18 to 22 who worked more than 35 weeks in the year and had usual weekly hours of more than 35. The wage ratio is RMC divided by the median wage (the wage at the 50th percentile of the wage distribution). Table A.3 shows the raw ratio, as well as the ratio predicted by fitting a line to the raw values. In our regression analysis, we used the raw values.

Table A.3
Regular Military Compensation/Wage Ratio for High School Graduates, Ages 18–22

Year	Males		Females	
	Raw	Predicted	Raw	Predicted
1999	1.36	1.49	1.54	1.62
2000	1.38	1.52	1.68	1.67
2001	1.53	1.55	1.61	1.72
2002	1.58	1.59	1.72	1.77
2003	1.76	1.62	1.86	1.82
2004	1.70	1.65	2.10	1.87
2005	1.64	1.68	1.90	1.92
2006	1.76	1.71	1.93	1.97
2007	1.82	1.75	2.09	2.02
2008	1.72	1.78	2.04	2.07
2009	1.81	1.81	2.15	2.12
2010	2.02	1.84	2.31	2.17
2011	2.13	1.88	2.25	2.22
2012	2.03	1.91	2.22	2.27
2013	1.85	1.94	2.17	2.32
2014	1.82	1.97	2.55	2.36
2015	2.07	2.00	2.36	2.41
2016	1.88	2.04	2.15	2.46
2017	1.95	2.07	2.68	2.51

SOURCES: Directorate of Compensation, 1999–2017; U.S. Census Bureau, 2018.

NOTE: RMC is for an E-4 with four YOS. The wage ratio is RMC divided by the median wage for 18- to 22-year-old workers with high school (and not additional) education who worked more than 35 weeks in the year and had more than 35 usual weekly hours of work.

Recruiting Rates for Armed Forces Qualification Test Categories I–IIIB and Regression Estimates

Recruiting Rates

The recruiting rate is the ratio of NPS HSDG enlisted accessions to the population of high school completers, net those who went on to complete four or more years of college. Accession data are from the military enlistment processing station (MEPS) file. Data on high school completers and on the percentage of high school completers who had completed four or more years of college by ages 25 to 29 are from NCES. We calculated recruiting rates by AFQT category. The category is given directly in MEPS data, but it is not present in NCES data. To allocate our adjusted high school completer population by AFQT category, we used the 1997 National Longitudinal Survey of Youth (NLSY), which administered the AFQT to a representative sample of young adults.

NCES provides data on recent high school completers, by gender, for 1960 through 2017 (NCES, 2018a). We also drew on NCES data to calculate the percentage of 25- to 29-year-olds who completed bachelor's degrees or higher conditional on completing high school or higher (NCES, 2018b). We assumed a modal age of 18 for high school completers and a modal age of 27 for the 25- to 29-year-olds who completed bachelor's degrees or higher (i.e., nine years later: age 27 minus age 18). The completion-rate data were available through 2017. We fitted linear trend models to the higher-degree (bachelor's or higher) completion data and used the estimated trend models to predict the higher-degree completion rates for high school completers for the span covered by our data, 1999 to 2017. We deducted these percentages from the population of completers. What remained was the number of high school completers not expected to later complete four-year degrees or higher. We assumed that this was the population to be recruited into the military.

The 1997 NLSY is the most recent renorming of the ASVAB (from which the AFQT score is calculated). NLSY provides information on the AFQT score distribution for 18- to 23-year-old men and women. We used the percentage of the population in each category to estimate the percentage of our net high school completer population by category. This did not account for the possibility that the AFQT distribution conditional on high school completion differs from that of the unconditional population. The high school completion rate in 2000 was 87 percent for men and 89 percent for women (NCES, 2018b), suggesting that the AFQT distribution for high school completers is likely to be close to that for the 18- to 23-year-old population overall.

Table B.1 shows the total number of male and female high school completers, the percentage of completers expected to complete four or more years of college, and the number of high school completers net of the latter. Tables B.2 through B.5 show the recruiting rates by service and gender for AFQT categories I, II, IIIA, and IIIB.

Table B.1
High School Completers, 1999–2018, Net Those Predicted to Complete Four
or More Years of College

Calendar Year	High School Completers, in Thousands		Percentage Predicted to Complete Bachelor's or Higher		High School Completers Net of Predicted Bachelor's or Higher Completers, in Thousands	
	Male	Female	Male	Female	Male	Female
1999	1,474	1,423	31.1	38.1	1,015	881
2000	1,251	1,505	31.7	38.6	855	924
2001	1,277	1,273	32.2	39.1	866	774
2002	1,412	1,384	32.7	39.7	950	835
2003	1,306	1,372	33.2	40.2	872	820
2004	1,327	1,425	33.8	40.8	879	844
2005	1,262	1,414	34.3	41.3	829	830
2006	1,328	1,363	34.8	41.8	866	793
2007	1,511	1,444	35.4	42.4	977	832
2008	1,640	1,511	35.9	42.9	1,052	862
2009	1,407	1,531	36.4	43.5	894	865
2010	1,679	1,482	36.9	44.0	1,059	830
2011	1,611	1,468	37.5	44.5	1,007	814
2012	1,622	1,581	38.0	45.1	1,006	868
2013	1,524	1,453	38.5	45.6	937	790
2014	1,423	1,445	39.0	46.1	867	778
2015	1,448	1,516	39.6	46.7	875	808
2016	1,517	1,620	40.1	47.2	908	855
2017	1,345	1,525	40.6	47.8	799	796
2018[a]	1,548	1,555	41.2	48.3	911	804

SOURCE: NCES, 2018.

NOTE: We fitted a linear trend to the percentage of 25- to 29-year-olds who completed bachelor's degrees or higher, conditional on completing high school or more, for 2005 through 2018. We assumed a median age of 27 for the 25- to 29-year-olds and a median age of 18 for high school completers, a nine-year difference, then used the linear trend to predict the percentage of high school completers in 1999 through 2018 who would complete bachelor's degrees or higher by nine years later.

[a] The number of high school completers in 2018 was predicted from a linear trend model fitted to high school completers in 1999 through 2017.

Table B.2
Army Recruiting Rates, 2000–2018: Armed Forces Qualification Test
Categories I, II, IIIA, and IIIB, as Percentages

	I		II		IIIA		IIIB	
Year	%Men	%Women	%Men	%Women	%Men	%Women	%Men	%Women
2000	2.7	0.4	5.3	1.3	8.1	2.2	8.3	2.6
2001	2.8	0.5	5.3	1.5	7.9	2.6	7.8	3.0
2002	3.5	0.5	5.3	1.4	7.1	2.2	6.6	2.3
2003	4.2	0.6	5.7	1.3	7.5	2.1	6.4	2.3
2004	5.5	0.7	7.2	1.5	9.4	2.3	7.9	2.4
2005	4.8	0.6	5.9	1.2	7.5	1.9	7.0	2.1
2006	4.0	0.5	5.1	1.1	5.9	1.6	6.6	2.1
2007	3.1	0.4	3.9	1.0	4.6	1.3	5.1	1.7
2008	3.1	0.4	4.1	1.0	4.9	1.3	5.7	1.9
2009	4.1	0.4	5.1	1.0	6.4	1.3	7.7	1.9
2010	4.4	0.6	5.5	1.2	6.7	1.7	8.5	2.5
2011	4.1	0.5	5.1	1.0	6.1	1.5	8.1	2.5
2012	3.3	0.4	4.9	0.9	6.4	1.3	7.9	2.0
2013	3.8	0.5	5.8	1.1	8.1	1.9	10.4	2.8
2014	3.0	0.4	4.8	0.9	7.1	1.7	8.7	2.3
2015	3.1	0.4	4.9	0.9	7.1	1.7	9.4	2.6
2016	3.0	0.4	4.7	0.9	6.5	1.5	8.9	2.5
2017	3.8	0.4	5.8	1.0	8.2	1.7	10.7	2.8
2018[a]	3.9	0.5	5.6	1.2	7.4	1.8	8.2	2.6

[a] Denominator is imputed.

Table B.3
Navy Recruiting Rates, 2000–2018: Armed Forces Qualification Test
Categories I, II, IIIA, and IIIB, as Percentages

Year	I %Men	I %Women	II %Men	II %Women	IIIA %Men	IIIA %Women	IIIB %Men	IIIB %Women
2000	2.2	0.3	4.2	0.9	5.4	1.5	7.4	1.7
2001	1.9	0.3	3.9	1.1	5.2	1.8	7.3	1.9
2002	2.1	0.3	3.4	0.9	4.4	1.4	5.7	1.4
2003	2.2	0.3	3.5	0.8	4.3	1.4	5.5	1.1
2004	3.1	0.4	4.4	1.0	5.1	1.5	5.5	1.1
2005	2.9	0.4	4.5	0.9	5.9	1.4	6.6	1.3
2006	2.8	0.4	4.1	1.0	4.8	1.4	3.9	1.3
2007	2.2	0.3	3.2	0.8	3.8	1.2	3.8	1.2
2008	2.3	0.4	3.3	0.9	3.8	1.2	3.4	1.3
2009	2.9	0.5	3.9	1.1	4.3	1.4	3.2	1.0
2010	2.8	0.5	3.5	1.2	3.8	1.8	1.9	0.8
2011	3.2	0.6	3.8	1.3	4.2	1.9	1.1	0.7
2012	3.4	0.6	4.2	1.3	4.7	2.0	1.1	0.7
2013	3.7	0.6	4.5	1.5	5.3	2.3	2.2	1.2
2014	3.6	0.5	4.3	1.3	5.1	2.1	1.4	0.8
2015	3.5	0.6	4.5	1.4	5.4	2.3	1.2	0.9
2016	3.0	0.4	3.5	1.0	4.2	1.7	1.3	0.9
2017	3.8	0.5	4.5	1.3	5.4	2.3	2.2	1.2
2018[a]	3.3	0.5	4.1	1.2	4.5	1.9	3.3	1.9

[a] Denominator is imputed.

Table B.4
Marine Corps Recruiting Rates, 2000–2018: Armed Forces Qualification Test
Categories I, II, IIIA, and IIIB, as Percentages

Year	I		II		IIIA		IIIB	
	%Men	%Women	%Men	%Women	%Men	%Women	%Men	%Women
2000	1.1	0.1	3.1	0.2	4.7	0.4	5.2	0.3
2001	1.1	0.1	3.1	0.3	4.8	0.5	5.2	0.4
2002	1.2	0.1	3.1	0.3	4.6	0.5	4.6	0.3
2003	1.5	0.1	3.5	0.3	4.7	0.5	4.5	0.3
2004	1.9	0.1	3.9	0.3	5.2	0.5	4.8	0.3
2005	2.0	0.1	4.2	0.3	5.7	0.5	5.8	0.5
2006	1.8	0.1	3.6	0.3	4.7	0.4	4.8	0.5
2007	1.7	0.1	3.4	0.3	4.3	0.4	4.6	0.5
2008	1.6	0.1	3.4	0.3	4.5	0.4	4.6	0.4
2009	1.7	0.1	3.5	0.3	4.6	0.5	4.4	0.4
2010	1.5	0.1	2.8	0.3	3.7	0.5	3.1	0.4
2011	1.6	0.1	3.3	0.3	4.4	0.5	3.2	0.4
2012	1.6	0.1	3.4	0.3	4.7	0.5	3.3	0.4
2013	1.7	0.2	3.7	0.4	5.2	0.6	3.9	0.5
2014	1.4	0.1	3.1	0.4	4.6	0.7	3.4	0.5
2015	1.5	0.1	3.5	0.4	5.4	0.7	3.9	0.4
2016	1.3	0.1	3.2	0.4	5.0	0.7	4.0	0.4
2017	1.8	0.2	4.1	0.4	6.1	0.8	4.9	0.5
2018[a]	1.6	0.2	3.5	0.5	4.9	0.7	4.3	0.6

[a] Denominator is imputed.

Table B.5
Air Force Recruiting Rates, 2000–2018: Armed Forces Qualification Test
Categories I, II, IIIA, and IIIB, as Percentages

Year	I %Men	I %Women	II %Men	II %Women	IIIA %Men	IIIA %Women	IIIB %Men	IIIB %Women
2000	1.5	0.2	3.6	1.0	4.7	1.7	3.5	1.4
2001	1.6	0.3	3.7	1.2	4.7	1.9	3.2	1.4
2002	1.8	0.3	3.7	1.2	4.6	1.9	3.0	1.4
2003	2.2	0.4	4.1	1.2	4.8	1.9	2.5	1.0
2004	2.9	0.5	4.8	1.4	5.4	2.0	2.7	1.1
2005	1.8	0.3	2.7	0.8	3.3	1.1	1.7	0.6
2006	2.3	0.4	3.8	1.2	4.2	1.7	2.5	1.2
2007	1.8	0.3	3.1	1.0	3.3	1.4	2.1	1.0
2008	1.8	0.3	2.9	0.9	3.1	1.3	1.9	0.9
2009	2.7	0.4	4.0	1.0	4.1	1.4	2.3	0.9
2010	2.4	0.4	3.4	1.0	3.4	1.2	0.9	0.4
2011	2.5	0.4	3.9	1.1	4.2	1.5	0.1	0.1
2012	2.6	0.4	3.9	1.0	4.2	1.4	0.2	0.1
2013	2.4	0.4	3.8	1.1	4.0	1.3	0.2	0.1
2014	2.4	0.4	3.6	1.0	3.8	1.4	0.5	0.2
2015	2.3	0.3	3.5	1.0	3.5	1.3	0.8	0.3
2016	2.7	0.4	4.0	1.2	4.1	1.6	1.6	0.6
2017	3.0	0.4	4.3	1.2	4.5	1.7	2.1	0.9
2018[a]	2.6	0.5	3.6	1.2	3.7	1.5	1.8	0.9

[a] Denominator is imputed.

Regression Estimates

Table B.6
Logit Regression of Recruiting Rate for Armed Forces Qualification Test Categories I–IIIB, Army and Navy

Variable	Army			Navy		
	Coefficient	Standard error	t	Coefficient	Standard error	t
Pay ratio	−0.2750	(0.153)	−1.801	0.2060	(0.153)	1.345
Pay ratio × Cat I	0.280	(0.167)	1.672	0.508	(0.168)	3.029
Pay ratio × Cat II	0.018	(0.126)	0.140	0.076	(0.115)	0.668
Pay ratio × Cat IIIB	−0.087	(0.212)	−0.412	−0.360	(0.388)	−0.928
Recruiting goal	0.006	(0.002)	2.695	0.012	(0.005)	2.356
Deployment	−0.012	(0.007)	−1.880	−0.105	(0.060)	−1.737
Unemployment	0.023	(0.010)	2.167	0.006	(0.014)	0.414
Cat I	−1.181	(0.329)	−3.592	−1.450	(0.315)	−4.607
Cat II	−0.329	(0.242)	−1.358	−0.326	(0.214)	−1.521
Cat IIIB	0.123	(0.348)	0.355	0.558	(0.661)	0.844
Female	−1.351	(0.079)	−16.992	−1.133	(0.063)	−17.998
Female × Cat I	−0.699	(0.097)	−7.220	−0.996	(0.078)	−12.833
Female × Cat II	−0.167	(0.087)	−1.928	−0.279	(0.071)	−3.916
Female × Cat IIIB	0.201	(0.104)	1.930	0.181	(0.144)	1.257
Post-2009	0.015	(0.064)	0.232	0.153	(0.063)	2.412
Post-2009 × Cat IIIB	0.360	(0.093)	3.882	−0.881	(0.156)	−5.633
Constant	−2.643	(0.353)	−7.495	−3.838	(0.428)	−8.977
R-squared	0.972			0.949		
Observations	144			144		
Pay ratio I	0.005	(0.167)	0.030	0.713	(0.202)	3.530
Pay ratio II	−0.257	(0.125)	−2.050	0.282	(0.139)	2.030
Pay ratio IIIA	−0.275	(0.153)	−1.801	0.206	(0.153)	1.345
Pay ratio IIIB	−0.362	(0.178)	−2.040	−0.154	(0.419)	−0.370

NOTE: Table shows the results of regressions of the logit recruiting rate on the given variables. Robust standard errors are shown. The reference category is category IIIA. The final three rows show the estimate for the pay ratio and the given category by combining the coefficient for pay ratio and the relevant interaction term.

Table B.7
**Logit Regression of Recruiting Rate for Armed Forces Qualification Test
Categories I–IIIB, Marine Corps and Air Force**

Variable	Marine Corps			Air Force		
	Coefficient	Standard error	t	Coefficient	Standard error	t
Pay ratio	0.3780	(0.122)	3.098	0.0760	(0.126)	0.600
Pay ratio × Cat I	0.090	(0.131)	0.685	0.712	(0.221)	3.226
Pay ratio × Cat II	−0.044	(0.113)	−0.391	0.256	(0.135)	1.899
Pay ratio × Cat IIIB	−0.014	(0.208)	−0.068	0.217	(0.655)	0.330
Recruiting goal	−0.010	(0.005)	−1.836	0.032	(0.005)	6.084
Deployment	0.007	(0.007)	1.057	−0.098	(0.099)	−0.987
Unemployment	−0.030	(0.008)	−3.509	−0.044	(0.031)	−1.438
Cat I	−1.340	(0.247)	−5.427	−1.901	(0.412)	−4.617
Cat II	−0.276	(0.202)	−1.362	−0.559	(0.244)	−2.286
Cat IIIB	−0.025	(0.361)	−0.069	−0.875	(1.090)	−0.803
Female	−2.363	(0.054)	−43.408	−1.033	(0.055)	−18.732
Female × Cat I	−0.347	(0.069)	−5.061	−1.022	(0.083)	−12.277
Female × Cat II	−0.103	(0.064)	−1.603	−0.333	(0.064)	−5.211
Female × Cat IIIB	−0.111	(0.094)	−1.186	0.029	(0.277)	0.104
Post-2009	0.003	(0.048)	0.059	−0.002	(0.084)	−0.024
Post-2009 × Cat IIIB	−0.161	(0.083)	−1.938	−1.614	(0.330)	−4.892
Constant	−3.201	(0.292)	−10.973	−3.877	(0.247)	−15.713
R-squared	0.990			0.895		
Observations	144			144		
Pay ratio I	0.468	(0.111)	4.220	0.788	(0.262)	3.010
Pay ratio II	0.334	(0.091)	3.680	0.331	(0.168)	1.980
Pay ratio IIIA	0.378	(0.122)	3.098	0.076	(0.126)	0.600
Pay ratio IIIB	0.364	(0.177)	2.050	0.292	(0.675)	0.430

NOTE: Table shows the results of regressions of the logit recruiting rate on the given variables. Robust standard are errors shown. The reference category is category IIIA. The final three rows show the estimate for the pay ratio and the given category by combining the coefficient for pay ratio and the relevant interaction term.

Table B.8
Logit Regression of Share of Non–High School Diploma Graduate Accessions for Armed Forces Qualification Test Categories I–IIIB, Army and Navy

Variable	Army			Navy		
	Coefficient	Standard error	t	Coefficient	Standard error	t
Pay ratio	0.3010	(0.232)	1.296	0.2900	(0.311)	0.930
Pay ratio × Cat I	1.043	(0.274)	3.810	1.623	(0.353)	4.597
Pay ratio × Cat II	0.428	(0.200)	2.140	0.428	(0.302)	1.417
Pay ratio × Cat IIIB	0.997	(0.650)	1.533	−0.102	(0.719)	−0.142
Recruiting goal	0.001	(0.008)	0.113	0.047	(0.008)	6.128
Deployment	0.033	(0.017)	1.916	−0.621	(0.137)	−4.523
Unemployment	−0.158	(0.030)	−5.268	−0.015	(0.029)	−0.513
Cat I	−3.480	(0.491)	−7.087	−4.045	(0.603)	−6.705
Cat II	−1.299	(0.371)	−3.501	−1.192	(0.511)	−2.332
Cat IIIB	−2.227	(1.029)	−2.165	−1.392	(1.195)	−1.165
Female	−0.801	(0.110)	−7.276	−0.652	(0.162)	−4.017
Female × Cat I	0.053	(0.154)	0.344	−0.304	(0.234)	−1.296
Female × Cat II	0.001	(0.135)	0.004	−0.013	(0.198)	−0.064
Female × Cat IIIB	−0.202	(0.329)	−0.614	0.227	(0.277)	0.818
Post-2009	−1.107	(0.134)	−8.248	−1.015	(0.140)	−7.262
Post-2009 × Cat IIIB	−0.703	(0.406)	−1.732	0.271	(0.382)	0.710
Constant	−0.773	(0.835)	−0.927	−3.718	(0.718)	−5.178
R-squared	0.849			0.799		
Observations	143			144		
Pay ratio I	1.344	(0.309)	4.340	1.912	(0.349)	5.480
Pay ratio II	0.729	(0.239)	3.050	0.718	(0.304)	2.360
Pay ratio IIIA	0.301	(0.232)	1.296	0.290	(0.311)	0.930
Pay ratio IIIB	1.298	(0.636)	2.040	0.187	(0.723)	0.260

NOTE: Table shows the results of regressions of the logit recruiting rate on the given variables. Robust standard errors are shown. The reference category is category IIIA. The final three rows show the estimate for the pay ratio and the given category by combining the coefficient for pay ratio and the relevant interaction term.

Table B.9
Logit Regression of Share of Non–High School Diploma Graduate Accessions for Armed Forces Qualification Test Categories I–IIIB, Marine Corps and Air Force

Variable	Marine Corps			Air Force		
	Coefficient	Standard error	t	Coefficient	Standard error	t
Pay ratio	0.5070	(0.367)	1.383	1.2990	(0.319)	4.076
Pay ratio × Cat I	1.823	(0.362)	5.033	−1.927	(0.399)	−4.834
Pay ratio × Cat II	0.282	(0.310)	0.908	−1.410	(0.345)	−4.083
Pay ratio × Cat IIIB	0.535	(0.470)	1.138	−0.062	(0.474)	−0.130
Recruiting goal	0.049	(0.014)	3.492	−0.014	(0.009)	−1.577
Deployment	−0.059	(0.015)	−3.931	−0.619	(0.111)	−5.586
Unemployment	−0.088	(0.025)	−3.439	0.274	(0.040)	6.895
Cat I	−3.787	(0.652)	−5.812	3.616	(0.737)	4.905
Cat II	−0.761	(0.549)	−1.385	2.948	(0.645)	4.570
Cat IIIB	−1.148	(0.776)	−1.480	0.066	(0.773)	0.085
Female	−0.685	(0.132)	−5.178	−0.534	(0.169)	−3.168
Female × Cat I	−0.197	(0.223)	−0.884	0.672	(0.210)	3.196
Female × Cat II	0.083	(0.165)	0.505	0.435	(0.191)	2.282
Female × Cat IIIB	−0.186	(0.213)	−0.871	−0.027	(0.270)	−0.099
Post-2009	−0.776	(0.176)	−4.410	0.244	(0.131)	1.859
Post-2009 × Cat IIIB	−0.535	(0.235)	−2.277	−0.001	(0.261)	−0.004
Constant	−4.388	(0.662)	−6.625	−7.244	(0.640)	−11.312
R-squared	0.708			0.721		
Observations	142			144		
Pay ratio I	2.330	(0.331)	7.040	−0.628	(0.371)	−1.690
Pay ratio II	0.789	(0.279)	2.830	−0.111	(0.317)	−0.350
Pay ratio IIIA	0.507	(0.367)	1.383	1.299	(0.319)	4.076
Pay ratio IIIB	1.042	(0.344)	3.030	1.237	(0.365)	3.390

NOTE: Table shows the results of regressions of the logit recruiting rate on the given variables. Robust standard are errors shown. The reference category is category IIIA. The final three rows show the estimate for the pay ratio and the given category by combining the coefficient for pay ratio and the relevant interaction term.

Additional Graphs Comparing Regular Military Compensation Percentiles in Least and Most Urban States

In Figures C.1 through C.4 we repeat versions of Figures 2.1b and 2.2, which show RMC by year of service compared with civilian wages. We use data from the Greenbooks and weight the CPS data based on the military gender mix. Figures C.1 and C.2 examine how enlisted RMC compares with civilian wages using high school graduates as the reference group for the first nine years of a career, those with some college as the reference group for those with 10–19 years of service, and those with an associate's degree for those enlisted with 20–30 years of service. Figure C.1 shows RMC compared with civilian wages for civilians who live in the least urban states. Figure C.2 shows RMC compared with civilian wages for civilians who live in the most urban states. RMC is a higher percentile of civilian wages compared with civilians who live in the most urban versus the least urban states for years one through nine.[1] This pattern flips as we compare RMC with civilians with more years of formal education as years of service increase. However, differences between the percentiles in the least and most urban states are relatively small for less-educated workers.

In Figures C.3 and C.4 we compare RMC for officers from the Greenbooks with civilians with bachelor's degrees (1–9 years of service) and master's degrees or more (10–30 years of service). While exhibiting a similar overall pattern to that found in Figure 2.2, officer RMC is at

---

[1] This may be due to more competition for low-skilled jobs in these areas, which leads to lower civilian wages; a full analysis is beyond the scope of this report.

a persistently and substantially higher percentage of civilian pay in the least urban states than in the most urban states; in most cases the difference is more than ten percentiles.

Figure C.1
Enlisted Regular Military Compensation, Civilian Wages for Civilians in the Least Urban States, and Regular Military Compensation Percentiles for Full-Time, Full-Year Workers with High School, Some College, or Associate's Degree, 2017

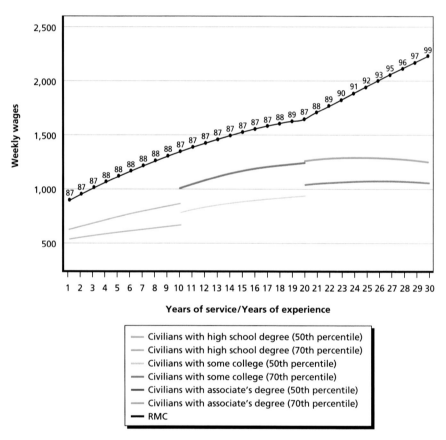

SOURCES: Directorate of Compensation, 2017; U.S. Census Bureau, 2018.
NOTE: RMC percentile varies by YOS (1–9 = high school, 10–19 = some college, and 20–30 = associate's degree). We weighted civilian-wage data by enlisted military gender mix. Colored lines are smoothed wage curves for the 50th and 70th percentiles of the given level of education. The black line is enlisted RMC; and the number above the black line is the percentile in the wage distribution for high school, some college, and associate's degree. Data are smoothed.

Figure C.2
Enlisted Regular Military Compensation, Civilian Wages for Civilians in the Most Urban States, and Regular Military Compensation Percentiles for Full-Time, Full-Year Workers with High School, Some College, or Associate's Degree, 2017

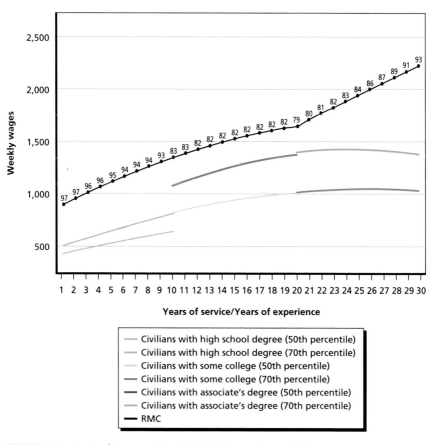

SOURCES: Directorate of Compensation, 2017; U.S. Census Bureau, 2018.
NOTE: RMC percentile varies by YOS (1–9 = high school, 10–19 = some college, and 20–30 = associate's degree). We weighted civilian-wage data by enlisted military gender mix. Colored lines are smoothed wage curves for the 50th and 70th percentiles of the given level of education. The black line is enlisted RMC; and the number above the black line is the percentile in the wage distribution for high school, some college, and associate's degree. Data are smoothed.

Figure C.3
Officer Regular Military Compensation, Civilian Wages for Civilians in the Least Urban States, and Regular Military Compensation Percentiles for Full-Time, Full-Year Workers with Bachelor's Degree or with Master's Degree or Higher, 2017

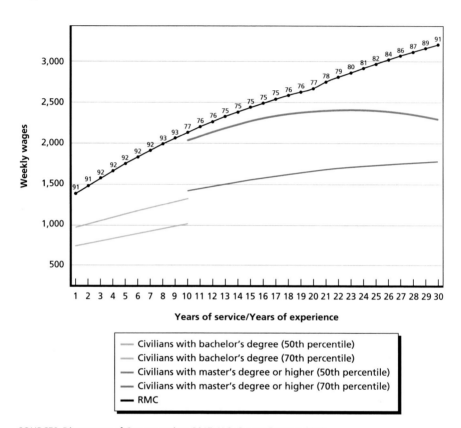

SOURCES: Directorate of Compensation, 2017; U.S. Census Bureau, 2018.
NOTE: RMC percentile varies by YOS (1–9 = bachelor's degree, 10–30 = master's degree or higher). We weighted civilian-wage data by military gender mix. Colored lines are smoothed wage curves for the 50th and 70th percentiles of the given level of education. The black line is enlisted RMC, and the numbers above the black line are the percentile in the wage distribution for a bachelor's degree and for a master's degree or higher. Data are smoothed.

Figure C.4
Officer Regular Military Compensation, Civilian Wages for Civilians in the Most Urban States, and Regular Military Compensation Percentiles for Full-Time, Full-Year Workers with Bachelor's Degree or with Master's Degree or Higher, 2017

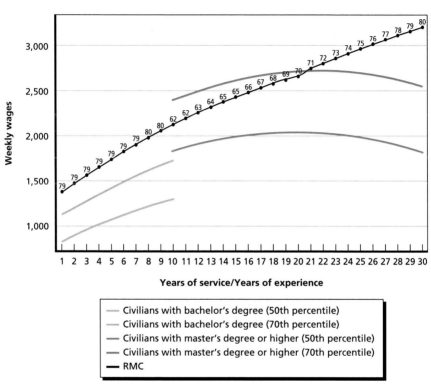

SOURCES: Directorate of Compensation, 2017; U.S. Census Bureau, 2018.
NOTE: RMC percentile varies by YOS (1–9 = bachelor's degree, 10–30 = master's degree or higher). We weighted civilian-wage data by military gender mix. Colored lines are smoothed wage curves for the 50th and 70th percentiles of the given level of education. The black line is enlisted RMC, and the numbers above the black line are the percentile in the wage distribution for a bachelor's degree and for a master's degree or higher. Data are smoothed.

Bibliography

"2001 US Military Basic Pay Charts," Navy CyberSpace, undated. As of July 8, 2019:
https://www.navycs.com/charts/2001-military-pay-chart.html

"2016: Another Lukewarm Year for Military Compensation," *Military Times*, January 18, 2016. As of July 31, 2018:
https://www.militarytimes.com/pay-benefits/military-pay-center/2016/01/18/2016-another-lukewarm-year-for-military-compensation/

Accession Policy Directorate, Office of the Under Secretary of Defense for Personnel and Readiness, "Accession Goals, Achievement and Quality," 2017, not available to the general public.

Acemoglu, Daron, and David Autor, "Skills, Tasks and Technologies: Implications for Employment and Earnings," in Orley Ashenfelter and David Card, eds., *Handbook of Labor Economics*, Vol. 4B, Amsterdam: North Holland/Elsevier, 2011, pp. 1043–1171.

Acemoglu, Daron, and Pascual Restrepo, "Robots and Jobs: Evidence from US Labor Markets," Cambridge, Mass.: National Bureau of Economic Research (NBER), Working Paper No. 23285, 2017.

———, "The Race between Man and Machine: Implications of Technology for Growth, Factor Shares, and Employment." *American Economic Review*, Vol. 108, No. 6, 2018, pp. 1488–1542.

Air Force Institute of Technology, "About the Graduate School," July 27, 2018. As of July 31, 2018:
https://www.afit.edu/EN/page.cfm?page=134

Alabdulkareem, Ahmad, Morgan R. Frank, Lijun Sun, Bedoor AlShebli, César Hidalgo, and Iyad Rahwan, "Unpacking the Polarization of Workplace Skills," *Science Advances*, Vol. 4, No. 7, 2018, eaao6030.

Asch, Beth J., Paul Heaton, James Hosek, Paco Martorell, Curtis Simon, and John T. Warner, *Cash Incentives and Military Enlistment, Attrition, and Reenlistment*, Santa Monica, Calif.: RAND Corporation, MG-950-OSD, 2010. As of May 19, 2017:
https://www.rand.org/pubs/monographs/MG950.html

Asch, Beth J., James Hosek, and John T. Warner, *An Analysis of Pay for Enlisted Personnel*, Santa Monica, Calif.: RAND Corporation, DB-344-OSD, 2001. As of March 6, 2018:
https://www.rand.org/pubs/documented_briefings/DB344.html

———, "New Economics of Manpower in the Post–Cold War Era," in Todd Sandler and Keith Hartley, eds., *Handbook of Defense Economics*, Vol. 2, New York: Elsevier, 2007, pp. 1075–1138.

Asch, Beth J., John A. Romley, and Mark E. Totten, *The Quality of Personnel in the Enlisted Ranks*, Santa Monica, Calif.: RAND Corporation, MG-324-OSD, 2005. As of November 23, 2017:
https://www.rand.org/pubs/monographs/MG324.html

Austin, Benjamin A., Edward L. Glaeser, and Lawrence H. Summers, "Jobs for the Heartland: Place-Based Policies in 21st Century America," Cambridge, Mass.: NBER Working Paper No. 24548, April 2018.

Autor, David, "Why Are There Still So Many Jobs? The History and Future of Workplace Automation," *Journal of Economic Perspectives*, Vol. 29, No. 3, 2015, pp. 3–30.

———, "Work of the Past, Work of the Future," Richard T. Ely Lecture, *American Economic Association: Papers and Proceedings*, Vol. 109, No. 5, May 2019, pp. 1–32. As of September 30, 2019:
https://economics.mit.edu/files/16724

Autor, David H., and David Dorn, "The Growth of Low-Skill Service Jobs and the Polarization of the US Labor Market," *American Economic Review*, Vol. 103, No. 5, 2013, pp. 1553–1597.

Autor, David H., Lawrence F. Katz, and Melissa S. Kearney, "The Polarization of the U.S. Labor Market," *American Economic Review*, Vol. 96, No. 2, 2006, pp. 189–194.

Autor, David H., Frank Levy, and Richard J. Murnane, "The Skill Content of Recent Technological Change: An Empirical Exploration," *Quarterly Journal of Economics*, Vol. 118, No. 4, 2003, pp. 1279–1333.

Bedno, Sheryl A., Christine E. Lang, William E. Daniell, Andrew R. Wiesen, Bennett Datu, and David W. Niebuhr, "Association of Weight at Enlistment with Enrollment in the Army Weight Control Program and Subsequent Attrition in the Assessment of Recruit Motivation and Strength Study," *Military Medicine*, Vol. 175, No. 3, March 2010, pp. 188–193. As of August 28, 2018:
http://www.amsara.amedd.army.mil/Documents/ARMS_Publication/2.%20Bedno-Mar-2010.pdf

Buddin, Richard, *Analysis of Early Military Attrition Behavior*, Santa Monica, Calif.: RAND Corporation, R-3069-MIL, 1984. As of November 10, 2017:
https://www.rand.org/pubs/reports/R3069.html

————, *Success of First-Term Soldiers: The Effects of Recruiting Practices and Recruit Characteristics*, Santa Monica, Calif.: RAND Corporation, MG-262-A, 2005. As of December 21, 2017:
https://www.rand.org/pubs/monographs/MG262.html

Bureau of Labor Statistics, U.S. Department of Labor, "Databases, Tables, and Calculators by Subject," undated. As of August 30, 2018:
https://www.bls.gov/data/

Chapman, Chris, Jennifer Laird, Nicole Ifill, and Angelina KewalRamani, *Trends in High School Dropout and Completion Rates in the United States: 1972–2009—Compendium Report*, Washington, D.C.: U.S. Department of Education, National Center for Education Statistics, Institute of Education Sciences, NCES 2012-006, October 2011. As of January 13, 2018:
https://nces.ed.gov/pubs2012/2012006.pdf

Chetty, Raj, and Nathaniel Hendren, "The Effects of Neighborhoods on Intergenerational Mobility I: Childhood Exposure Effects," *Quarterly Journal of Economics*, Vol. 133, No. 3, 2018a, pp. 1107–1162.

————, "The Effects of Neighborhoods on Intergenerational Mobility II: County Level Estimates," *Quarterly Journal of Economics*, Vol. 133, No. 3, 2018, pp. 1663–1228.

Chetty, Raj, Nathaniel Hendren, and Lawrence Katz, "The Effects of Exposure to Better Neighborhoods on Children: New Evidence from the Moving to Opportunity Experiment," *American Economic Review*, Vol. 106, No. 4, 2016, pp. 855–902.

Christensen, Garret, "Occupational Fatalities and the Labor Supply: Evidence from the Wars in Iraq and Afghanistan," *Journal of Economic Behavior and Organization*, Vol. 139, July 2017, pp. 182–195.

Clarke, Margaret Z., "Geographic Deregulation of Banking and Economic Growth," *Journal of Money, Credit and Banking*, Vol. 36, No. 5, 2004, pp. 929–942.

CNA Analysis and Solutions, "Population Representation in the Military Services," undated. As of May 19, 2017:
https://www.cna.org/research/pop-rep

Currie, Janet, and W. Bentley MacLeod, "First Do No Harm? Tort Reform and Birth Outcomes," *Quarterly Journal of Economics*, Vol. 123, No. 2, May 2008, pp. 795–830.

Defense Travel Management Office (DTMO), "Basic Allowance for Housing (BAH) Frequently Asked Questions," Alexandria, Va.: DTMO, January 27, 2015.

————, *A Primer on the Basic Allowance for Housing (BAH) for the Uniformed Services*, Alexandria, Va.: DTMO, January 2018.

Dertouzos, James N., *Recruiter Incentives and Enlistment Supply*, Santa Monica, Calif.: RAND Corporation, R-3065-MIL, 1985. As of May 19, 2017:
http://www.rand.org/pubs/reports/R3065.html

Directorate of Compensation, Office of the Under Secretary of Defense for Personnel and Readiness, *Selected Military Compensation Tables*, Washington, D.C.: U.S. Department of Defense, 1999.

———, *Selected Military Compensation Tables*, Washington, D.C.: U.S. Department of Defense, 2000.

———, *Selected Military Compensation Tables*, Washington, D.C.: U.S. Department of Defense, 2001.

———, *Selected Military Compensation Tables*, Washington, D.C.: U.S. Department of Defense, 2002.

———, *Selected Military Compensation Tables*, Washington, D.C.: U.S. Department of Defense, 2003. As of August 5, 2018: https://militarypay.defense.gov/Portals/3/Documents/Reports/greenbook_fy2003.pdf

———, *Selected Military Compensation Tables*, Washington, D.C.: U.S. Department of Defense, January 1, 2004. As of August 5, 2018: https://militarypay.defense.gov/Portals/3/Documents/Reports/greenbook_fy04.pdf

———, *Selected Military Compensation Tables*, Washington, D.C.: U.S. Department of Defense, January 1, 2005. As of August 5, 2018: https://militarypay.defense.gov/Portals/3/Documents/Reports/greenbook_fy05.pdf

———, *Selected Military Compensation Tables*, Washington, D.C.: U.S. Department of Defense, January 1, 2006. As of August 5, 2018: https://militarypay.defense.gov/Portals/3/Documents/Reports/greenbook2_fy06.pdf

———, *Selected Military Compensation Tables*, Washington, D.C.: U.S. Department of Defense, April 1, 2007. As of August 5, 2018: https://militarypay.defense.gov/Portals/3/Documents/Reports/GreenBook_APRIL_40YOS_2007_Dist.pdf

———, *Selected Military Compensation Tables*, Washington, D.C.: U.S. Department of Defense, January 1, 2008. As of August 5, 2018: https://militarypay.defense.gov/Portals/3/Documents/Reports/GreenBook_2008.pdf

———, *Selected Military Compensation Tables*, Washington, D.C.: U.S. Department of Defense, January 1, 2009. As of August 5, 2018: https://militarypay.defense.gov/Portals/3/Documents/Reports/GreenBook_2009.pdf

———, *Selected Military Compensation Tables*, Washington, D.C.: U.S. Department of Defense, January 1, 2010. As of August 5, 2018: https://militarypay.defense.gov/Portals/3/Documents/Reports/GreenBook_2010.pdf

———, *Selected Military Compensation Tables*, Washington, D.C.: U.S. Department of Defense, January 1, 2011. As of August 5, 2018: https://militarypay.defense.gov/Portals/3/Documents/Reports/GreenBook_2011.pdf

———, *Selected Military Compensation Tables*, Washington, D.C.: U.S. Department of Defense, January 1, 2012. As of August 5, 2018: https://militarypay.defense.gov/Portals/3/Documents/Reports/GreenBook_2012.pdf

———, *Selected Military Compensation Tables*, Washington, D.C.:
U.S. Department of Defense, January 1, 2013. As of August 5, 2018:
https://militarypay.defense.gov/Portals/3/Documents/Reports/GreenBook_2013.pdf

———, *Selected Military Compensation Tables*, Washington, D.C.:
U.S. Department of Defense, January 1, 2014. As of August 5, 2018:
https://militarypay.defense.gov/Portals/3/Documents/Reports/GreenBook_2014.pdf

———, *Selected Military Compensation Tables*, Washington, D.C.:
U.S. Department of Defense, January 1, 2015. As of August 5, 2018:
https://militarypay.defense.gov/Portals/3/Documents/Reports/GreenBook_2015.pdf

———, *Selected Military Compensation Tables*, Washington, D.C.:
U.S. Department of Defense, January 1, 2016. As of August 4, 2018:
https://militarypay.defense.gov/Portals/3/Documents/Reports/GreenBook_2016.pdf

———, *Selected Military Compensation Tables*, Washington, D.C.:
U.S. Department of Defense, January 1, 2017. As of December 30, 2017:
http://militarypay.defense.gov/Portals/3/Documents/Reports/GreenBook_2017
.pdf?ver=2017-04-24-232444-753

DoD—*See* U.S. Department of Defense

Drasgow, Fritz, *Tailored Adaptive Personality Assessment System (TAPAS)*, briefing
presented to the International Personnel Assessment Council, July 22, 2013. As of
August 28, 2018:
http://www.ipacweb.org/Resources/Documents/conf13/drasgow.pdf

Eighmey, John, "Why Do Youth Enlist? Identification of Underlying Themes,"
Armed Forces and Society, Vol. 32, No. 2, 2006, pp. 307–328.

Federal Reserve Bank of St. Louis, "Employment Cost Index: Wages and Salaries—
Private Industry Workers," undated a. As of July 31, 2018:
https://fred.stlouisfed.org/series/ECIWAG

———, "Real Median Household Income in the United States," undated b. As of
July 31, 2018:
https://fred.stlouisfed.org/series/MEHOINUSA672N

Ganong, Peter, and Daniel Shoag, "Why Has Regional Income Convergence in the
U.S. Declined?" *Journal of Urban Economics*, Vol. 102, November 2017, pp. 76–90.

Glaeser, Edward L., *Triumph of the City*, New York: Penguin Press, 2011.

———, ed., *Agglomeration Economics*, Chicago: University of Chicago Press, 2010.

Granger, C. W. J., and P. Newbold, "Spurious Regressions in Economics," *Journal
of Econometrics*, Vol. 2, No. 2, July 1974, pp. 111–120.

Heffner, Tonia S., Roy C. Campbell, and Fritz Drasgow, *Select for Success: A Toolset
for Enhancing Soldier Accessioning*, U.S. Army Research Institute for the Behavioral
and Social Sciences, Special Report 70, March 2011. As of August 28, 2018:
http://www.dtic.mil/dtic/tr/fulltext/u2/a554057.pdf

Holmes, Thomas J., "The Effect of State Policies on the Location of Manufacturing: Evidence from State Borders," *Journal of Political Economy*, Vol. 106, No. 4, August 1998, pp. 667–705.

Hosek, James, Beth J. Asch, Michael G. Mattock, and Troy D. Smith, *Military and Civilian Pay Levels, Trends, and Recruit Quality*, Santa Monica, Calif.: RAND Corporation, RR-2396-OSD, 2018. As of September 30, 2019:
https://www.rand.org/pubs/research_reports/RR2396.html

Hosek, James, and Paco Martorell, *How Have Deployments During the War on Terrorism Affected Reenlistment?* Santa Monica, Calif.: RAND Corporation, MG-873-OSD, 2009. As of November 23, 2017:
https://www.rand.org/pubs/monographs/MG873.html

Hosek, James, Christine E. Peterson, Jeannette Van Winkle, and Hui Wang, *A Civilian Wage Index for Defense Manpower*, Santa Monica, Calif.: RAND Corporation, R-4190-FMP, 1992. As of March 6, 2018:
https://www.rand.org/pubs/reports/R4190.html

Kapp, Lawrence, *Recruiting and Retention: An Overview of FY 2011 and FY 2012 Results for Active and Reserve Component Enlisted Personnel*, Congressional Research Service, RL32965, May 10, 2013. As of March 11, 2018:
https://fas.org/sgp/crs/natsec/RL32965.pdf

Kilburn, M. Rebecca, and Jacob Alex Klerman, *Enlistment Decisions in the 1990s: Evidence from Individual-Level Data*, Santa Monica, Calif.: RAND Corporation, MR-944-OSD/A, 1999. As of July 27, 2018:
https://www.rand.org/pubs/monograph_reports/MR944.html

Kleiner, Morris M., "Battling over Jobs: Occupational Licensing in Health Care," *American Economic Review Papers and Proceedings*, Vol. 106, No. 5, 2016, pp. 165–170.

Knapp, David, Beth J. Asch, Michael G. Mattock, and James Hosek, *An Enhanced Capability to Model How Compensation Policy Affects U.S. Department of Defense Civil Service Retention and Cost*, Santa Monica, Calif.: RAND Corporation, RR-1503-OSD, 2016. As of July 27, 2018:
https://www.rand.org/pubs/research_reports/RR1503.html

Knapp, David, Bruce R. Orvis, Christopher E. Maerzluft, and Tiffany Tsai, *Resources Required to Meet the U.S. Army's Enlisted Recruiting Requirements Under Alternative Recruiting Goals, Conditions, and Eligibility Policies*, Santa Monica, Calif.: RAND Corporation, RR-2364-A, 2018. As of July 27, 2018:
https://www.rand.org/pubs/research_reports/RR2364.html

Laurence, Janice H., *Education Standards for Military Enlistment and the Search for Successful Recruits*, Alexandria, Va.: Human Resources Research Organization, FR-PRD-84-4, February 1984. As of November 10, 2017:
http://www.dtic.mil/dtic/tr/fulltext/u2/a139718.pdf

Loughran, David S., and Bruce R. Orvis, *The Effect of the Assessment of Recruit Motivation and Strength (ARMS) Program on Army Accessions and Attrition*, Santa Monica, Calif.: RAND Corporation, TR-975-A, 2011. As of July 27, 2018:
https://www.rand.org/pubs/technical_reports/TR975.html

"Marine Corps University," Wikipedia, July 1, 2018. As of July 31, 2018:
https://en.wikipedia.org/wiki/Marine_Corps_University

McFadden, Daniel L., "Econometric Analysis of Qualitative Response Models," in Zvi Griliches and Michael D. Intriligator, eds., *Handbook of Econometrics*, Vol. 2, Amsterdam: Elsevier, 1983, pp. 1395–1457.

Murray, Carla Tighe, senior analyst for military compensation and health care, Congressional Budget Office, *Evaluating Military Compensation, Statement of Carla Tighe Murray Before the Subcommittee on Personnel, Committee on Armed Services, United States Senate*, Washington, D.C.: Congressional Budget Office, April 28, 2010. As of February 10, 2018:
https://www.cbo.gov/sites/default/files/111th-congress-2009-2010/reports/04-28-militarypay.pdf

Nataraj, Shanthi, M. Wade Markel, Jaime L. Hastings, Eric V. Larson, Jill Luoto, Christopher E. Maerzluft, Craig A. Myatt, Bruce R. Orvis, Christina Panis, Michael Powell, Jose Rodriguez, and Tiffany Tsai, *Evaluating the Army's Ability to Regenerate: History and Future Options*, Santa Monica, Calif.: RAND Corporation, RR-1637-A, 2017. As of May 19, 2017:
https://www.rand.org/pubs/research_reports/RR1637.html

National Center for Education Statistics, Institute of Education Sciences, "Number and Percentage Distribution of First-Time Postsecondary Students Starting at 2- and 4-Year Institutions During the 2011–12 Academic Year, by Attainment and Enrollment Status and Selected Characteristics: Spring 2014," *Digest of Education Statistics*, Table 326.50, February 2017. As of July 16, 2019:
https://nces.ed.gov/programs/digest/d17/tables/dt17_326.50.asp

———, "Recent High School Completers and Their Enrollment in College, by Sex and Level of Institution: 1960 through 2017," *Digest of Education Statistics*, Table 302.10, July 2018a. As of July 16, 2019:
https://nces.ed.gov/programs/digest/d18/tables/dt18_302.10.asp

———, "Percentage of Persons 25 to 29 Years Old with Selected Levels of Educational Attainment, by Race/Ethnicity and Sex: Selected Years, 1920 Through 2018," *Digest of Education Statistics*, Table 104.20, October 2018b. As of July 16, 2019:
https://nces.ed.gov/programs/digest/d18/tables/dt18_104.20.asp

Naval Postgraduate School, "Programs and Degrees," undated. As of July 31, 2018:
https://my.nps.edu/degree-programs

NCES—*See* National Center for Education Statistics

Nye, Christopher D., Fritz Dragow, Oleksandr S. Chernyshenko, Stephen Stark, U. Christean Kubisiak, Leonard A. White, and Irwin Jose, *Assessing the Tailored Adaptive Personality Assessment System (TAPAS) as an MOS Qualification Instrument*, Fort Belvoir, Va.: U.S. Army Research Institute for the Behavioral and Social Sciences, Technical Report 1312, August 2012. As of May 25, 2017: http://www.dtic.mil/dtic/tr/fulltext/u2/a566090.pdf

Office of the Deputy Assistant Secretary of Defense for Military Community and Family Policy, *2015 Demographics: Profile of the Military Community*, c. 2016. As of August 2, 2018: http://download.militaryonesource.mil/12038/MOS/Reports/2015-Demographics -Report.pdf

Office of People Analytics, U.S. Department of Defense, military enlistment processing station files provided to the authors, undated.

———, Status of Forces Surveys of Active Duty Members results provided to the authors, 2017. Data for 2009 and 2016 were provided in December 2017.

———, Status of Forces Surveys of Active Duty Members results provided to the authors, 2018. Data for 2017 were provided in October 2018.

———, "Summer 2017 Propensity Update: Youth Poll Study Findings," unpublished manuscript, February 2018.

Office of the Under Secretary of Defense (Comptroller), *Military Personnel Programs (M-1): Department of Defense Budget—March Budget Amendment to the Fiscal Year 2017 President's Budget Request for BASE + Overseas Contingency Operations (OCO)*, March 2017. As of July 31, 2018: http://comptroller.defense.gov/Portals/45/Documents/defbudget/fy2017/ marchAmendment/fy2017_m1a.pdf

Office of the Under Secretary of Defense for Personnel and Readiness,"Annual Pay Adjustment," *Military Compensation*, undated. As of July 31, 2018: http://militarypay.defense.gov/Pay/Basic-Pay/AnnualPayRaise/

———, *Report of the Seventh Quadrennial Review of Military Compensation*, Washington, D.C.: U.S. Department of Defense, 1992.

———, *Report of the Ninth Quadrennial Review of Military Compensation*, Vols. I–V, Washington, D.C.: U.S. Department of Defense, March 2002. As of July 31, 2018: http://militarypay.defense.gov/Portals/3/Documents/Reports/9th_QRMC_Report _Volumes_I_-_V.pdf

———, *Report of the Tenth Quadrennial Review of Military Compensation*, Vol. I: *Cash Compensation*, Washington, D.C.: U.S. Department of Defense, February 2008. As of August 28, 2018: https://militarypay.defense.gov/Portals/3/Documents/Reports/10th_QRMC_2008 _Vol_I_Cash_Compensation.pdf

———, *Population Representation in the Military Services: Fiscal Year 2010*, Washington, D.C.: U.S. Department of Defense, 2011. As of August 2, 2018: https://prhome.defense.gov/M-RA/Inside-M-RA/MPP/Accession-Policy/ Pop-Rep/2010/

———, *Report of the Eleventh Quadrennial Review of Military Compensation: Main Report*, Washington, D.C.: U.S. Department of Defense, June 2012a. As of July 31, 2018: http://militarypay.defense.gov/Portals/3/Documents/Reports/11th_QRMC_ Main_Report_FINAL.pdf?ver=2016-11-06-160559-590

———, *Report of the Eleventh Quadrennial Review of Military Compensation: Supporting Research Papers*, Washington, D.C.: U.S. Department of Defense, June 2012b. As of July 31, 2018: http://militarypay.defense.gov/Portals/3/Documents/Reports/11th_QRMC_ Supporting_Research_Papers_(932pp)_Linked.pdf

———, *Population Representation in the Military Services: Fiscal Year 2015*, Washington, D.C.: U.S. Department of Defense, c. 2016. As of August 2, 2018: https://prhome.defense.gov/M-RA/Inside-M-RA/MPP/Accession-Policy/Pop-Rep/ 2015/

———, *Military Compensation Background Papers: Compensation Elements and Related Manpower Cost Items, Their Purposes and Legislative Backgrounds*, 8th Edition, Washington, D.C.: U.S. Department of Defense, 2018. As of January 31, 2019: https://militarypay.defense.gov/Portals/3/Documents/Reports/Mil-Comp_8thEdition .pdf?ver=2018-09-01-181142-307

Orvis, Bruce R., Michael Childress, and J. Michael Polich, *Effect of Personnel Quality on the Performance of Patriot Air Defense System Operators*, Santa Monica, Calif.: RAND Corporation, R-3901-A, 1992. As of November 23, 2017: https://www.rand.org/pubs/reports/R3901.html

Orvis, Bruce R., Steven Garber, Philip Hall-Partyka, Christopher E. Maerzluft, and Tiffany Tsai, *Recruiting Strategies to Support the Army's All-Volunteer Force*, Santa Monica, Calif.: RAND Corporation, RR-1211-A, 2016. As of November 23, 2017: https://www.rand.org/pubs/research_reports/RR1211.html

Orvis, Bruce R., Christopher E. Maerzluft, Sung-Bou Kim, Michael G. Shanley, and Heather Krull, *Prospective Outcome Assessment for Alternative Recruit Selection Policies*, Santa Monica, Calif.: RAND Corporation, RR-2267-A, 2018. As of August 28, 2018: https://www.rand.org/pubs/research_reports/RR2267.html

"Project A: The U.S. Army Selection and Classification Project," special issue, *Personnel Psychology*, Vol. 43, No. 2, 1990, pp. 231–379. As of September 29, 2019: https://onlinelibrary.wiley.com/toc/17446570/1990/43/2

Public Law 110-252, Supplemental Appropriations Act, 2008, June 30, 2008. As of August 5, 2018: https://www.gpo.gov/fdsys/pkg/PLAW-110publ252/content-detail.html

Rubenstein, Yona, and Yoram Weiss, "Post Schooling Wage Growth: Investment, Search and Learning," in Erik Hanushek and Finis Welch, eds., *Handbook of the Economics of Education*, Vol. 1, Amsterdam: Elsevier North Holland, 2007, pp. 1–68.

Schmitz, Edward J., and Michael J. Moskowitz, with David Gregory and David L. Reese, *Recruiting Budgets, Recruit Quality, and Enlisted Performance*, Alexandria, Va.: Center for Naval Analyses, CRM D0017035.A2/Final, February 2008. As of July 31, 2018: https://www.cna.org/CNA_files/PDF/d0017035.a2.pdf

Scribner, Barry L., D. Alton Smith, Robert H. Baldwin, and Robert L. Phillips, "Are Smart Tankers Better? AFQT and Military A Productivity," *Armed Forces and Society*, Vol. 12, No. 2, 1986, pp. 193–206.

Segall, Daniel O., *Development and Evaluation of the 1997 ASVAB Score Scale*, Alexandria, Va.: Defense Manpower Data Center, July 2004. As of July 27, 2018: http://official-asvab.com/docs/1997score_scale.pdf

Sellman, W. Steven, "Predicting Readiness for Military Service: How Enlistment Standards Are Established," draft prepared for the National Assessment Governing Board, September 30, 2004. As of August 1, 2018: https://www.nagb.gov/content/nagb/assets/documents/what-we-do/sellman.doc

Simon, Curtis J., and John T. Warner, "Managing the All-Volunteer Force in Time of War," *Economics of Peace and Security Journal*, Vol. 2, No. 1, January 2007, pp. 20–29.

Stark, Stephen, Oleksandr S. Chernyshenko, Fritz Drasgow, Christopher D. Nye, Leonard A. White, Tonia Heffner, and William L. Farmer, "From ABLE to TAPAS: A New Generation of Personality Tests to Support Military Selection and Classification Decisions," *Military Psychology*, Vol. 26, No. 3, 2014, pp. 153–164.

U.S. Army War College, "Military Education Level 1 Programs," undated. As of July 31, 2018: https://www.armywarcollege.edu/programs/mel_1.cfm

U.S. Census Bureau, Decennial Census, 1900–2010, "Percentage of the Total Population in Urban Areas," Iowa State University, Iowa Community Indicators Program, 2010. As of October 18, 2019: https://www.icip.iastate.edu/tables/population/urban-pct-states

———, "Current Population Survey (CPS): Data," March 6, 2018. As of August 31, 2018: https://www.census.gov/programs-surveys/cps/data-detail.html

U.S. Code, Title 26, Internal Revenue Code, Subtitle C, Employment Taxes, Chapter 21, Federal Insurance Contributions Act, January 3, 2012. As of August 2, 2018:
https://www.gpo.gov/fdsys/granule/USCODE-2011-title26/USCODE-2011
-title26-subtitleC-chap21/content-detail.html

U.S. Department of Defense, *Career Compensation for the Uniformed Forces: Army, Navy, Air Force, Marine Corps, Coast Guard, Coast and Geodetic Survey, Public Health Service: A Report and Recommendation for the Secretary of Defense by the Advisory Commission on Service Pay*, Washington, D.C.: U.S. Department of Defense, 1948.

———, "Fall 2016 Propensity Update, Youth Propensity: Youth Poll Study Findings," Joint Advertising and Marketing Research and Studies, Office of People Analytics, 2017, unpublished.

U.S. Government Accountability Office, *Military Personnel: Military and Civilian Pay Comparisons Present Challenges and Are One of Many Tools in Assessing Compensation*, Washington, D.C., GAO-10-561R, April 1, 2010. As of August 30, 2018:
https://www.gao.gov/products/GAO-10-561R

U.S. Naval War College, "Programs Offered," undated. As of July 31, 2018:
https://usnwc.edu/Academics-and-Programs/Programs-Offered

Warner, John, Curtis Simon, and Deborah Payne, "The Military Recruiting Productivity Slowdown: The Roles of Resources, Opportunity Cost and the Tastes of Youth," *Defence and Peace Economics*, Vol. 14, No. 5, October 2003, pp. 329–342.

Wenger, Jennie W., Zachary T. Miller, and Seema Sayala, *Recruiting in the 21st Century: Technical Aptitude and the Navy's Requirements*, Washington, D.C.: CNA Analysis and Solutions, CRM D0022305.A2/Final, May 2010. As of May 19, 2017:
https://www.cna.org/CNA_files/PDF/D0022305.A2.pdf

Winkler, John D., Judith C. Fernandez, and J. Michael Polich, *Effect of Aptitude on the Performance of Army Communications Operators*, Santa Monica, Calif.: RAND Corporation, R-4143-A, 1992. As of November 23, 2017:
https://www.rand.org/pubs/reports/R4143.html

Yule, G. Udny, "Why Do We Sometimes Get Nonsense Correlations Between Time-Series? A Study in Sampling and the Nature of Time-Series," *Journal of the Royal Statistical Society*, Vol. 89, No. 1, January 1926, pp. 1–63.